Biomechatronic Design
in Biotechnology

Biomechatronic Design in Biotechnology

A Methodology for Development of Biotechnological Products

Carl-Fredrik Mandenius
Professor in Engineering Biology
Linköping University, Sweden

Mats Björkman
Professor in Assembly Technology
Linköping University, Sweden

Published by John Wiley & Sons, Inc., Hoboken, New Jersey.
Published simultaneously in Canada.

For general information on our other products and services or for technical support, please contact our Customer Care Department within the United States at (800) 762-2974, outside the United States at (317) 572-3993 or fax (317) 572-4002.

Wiley also publishes its books in a variety of electronic formats and by print-on-demand. Not all content that is available in standard print versions of this book may appear or be packaged in all book formats. If you have purchased a version of this book that did not include media that is referenced by or accompanies a standard print version, you may request this media by visiting *http://booksupport.wiley.com*. For more information about Wiley products, visit us *www.wiley.com*.

Library of Congress Cataloging-in-Publication Data:

Mandenius, Carl-Fredrik, 1954-
Biomechatronic design in biotechnology : a methodology for development of biotechnological products / Carl-Fredrik Mandenius, Mats Björkman.
p. ; cm.
Includes bibliographical references and index.
ISBN 978-0-470-57334-1 (cloth)
1. Biotechnology. 2. Biomechanics. 3. Biomedical engineering. I. Björkman, Mats, 1955- II. Title.
[DNLM: 1. Biotechnology–methods. 2. Biomedical Engineering–methods. 3. Chemistry Techniques, Analytical. W 82]
TP248.2.M366 2011
610.28–dc22

2010054066

oBook ISBN: 9781118067147
ePDF ISBN: 9781118067123
ePub ISBN: 9781118067130

10 9 8 7 6 5 4 3 2 1

Contents

Preface

The purpose of this book is to provide the reader with an introduction to systematic design principles and methodology when applied to biotechnology products. Certainly, none of these fields is new on the block – it is the combination of them that brings about a novel approach in this book. The theory of systematic design has almost entirely been devoted to mechanics and electronics, and the biotechnology field has had much of its roots in white biology and in (bio)chemical engineering.

Thus, we are dealing with a subject that lies on the border between biological technology and mechanical and electric engineering. The aim is to integrate important aspects of biological technology with mechanical and electric engineering. In writing a book of this type, there are two major ways of organizing the material, either from the perspective of mechanical design engineering or from the perspective of biotechnology. We have chosen the first for the simple reason that we have used mechatronics methodology from mechanical design engineering as a basis and applied it to biotechnology. When doing so, we have adapted the methodology to what we call a *biomechatronics* approach.

We presume the book will have mainly two categories of readers, those with a background in biotechnology and related areas and those with a background in mechanics and electronics. We have tried to keep most of the text on a level where both categories of readers would be able to understand the subject.

When this has not been possible, due to space constraints, we have instead provided rather detailed lists of reference literature.

We realize that a great deal of the biotechnical details in the application cases in Chapters 5–13 are probably rather difficult to understand for a person with a background in mechanics and/or electronics. You would need a thorough knowledge of biotechnology in order to comprehend everything in these chapters. However, the main ideas of how to utilize and work with the presented biomechatronic design methodology are possible to understand when reading these chapters. It is not necessary to understand all the biotechnological details in order to have great benefit from these chapters.

We have provided in the book nine application cases from a rather diverse collection of biotechnology products, such as biosensors, analytical instrumentations, production equipment for cell culturing, and protein purification. Some of the products could be characterized more as systems products rather than discrete physical products, for example, PAT-based quality systems.

Readers with their own practical experiences could select from these application cases those relevant to their area of practice. Still, it is necessary to first grasp the general methodology approach and tools, especially explained in Chapter 4 and related to fundamental design theory in Chapter 2, before starting reading specific application cases.

The readers who wish to have a complete overview should of course go through most of the chapters.

We take a great pleasure in expressing thanks to our colleagues Dr. Micael Derelöv and Dr. Jonas Detterfelt for contributing many valuable ideas and suggestions, in particular, to the initial studies of the subject of the book. Valuable contributions on inquiries and interviews with developers and companies were made by Maria Uhr and Annika Perhammar.

We also thank our academic colleagues in engineering, medicine, and biophysics: Drs. Katrin Zeilinger, Jörg Gerlach, Bo Liedberg, Danny van Noort, and Ingemar Lundström.

We are also grateful to many biotechnology companies and their personnel who have shared their experiences and endeavors in the development of biomechatronic products. In particular, we would like to mention Drs. Stefan Löfås, Ulf Jönsson, at Biacore; Lasse Mörtsell at Belach AB; Stellan Lindberg at Hemocue AB; Stefan Nilsson and Johan Rydén at Noster AB; Dario Kriz at European Institute of Science AB; and colleagues at Cellartis AB.

We owe gratitude to Abbott, Agilent, Affymetrix, Biodot, Charité Universitätsmedizin Berlin, GE Healthcare, Hemocue, Johnson & Johnson, Kibion, KronLab Chromatography, LifeSpan, Roche Diagnostics, Q-Sense and Mr. Anders Sandelin for providing us figures and pictures.

We would like to thank the Swedish Agency for Innovation Systems (VINNOVA) and Linköping University for financially supporting our research on this topic.

We appreciate the support from John Wiley & Sons, Inc., Hoboken, in the production of this book.

Finally, we would like to express our sincere gratitude to all the skillful scientists and engineers who have contributed immensely to the development of design science in a variety of areas of technology. Without them, this book would not have been realized.

Carl-Fredrik Mandenius
Mats Björkman

Linköping
November 2010

1

Introduction

1.1 SCOPE OF DESIGN

Design is a concept with many aspects. So far, there exists no generally accepted definition of the concept. The word *design* has different meanings in different disciplines and fields. However, in general terms the verb design normally does refer to the process of planning, constructing, and creating a physical structure and functions of a physical artifact. Design can also refer to the process of creating the structure and functions of systems or services. In most cases, the concept of design is related to the development of new products.

Design is also characterized by having significant impact on most areas of human life. Almost all objects we interact with have gone through a design stage: the house we live in, the household machines for food preparation, and the vehicles that transport us to our office. Our mobile telephone is integrated into a complex communication network designed for optimal interconnection. The pills to cure the headache after work are a result of drug design. The examples from daily life are endless.

Biomechatronic Design in Biotechnology: A Methodology for Development of Biotechnological Products, First Edition. Carl-Fredrik Mandenius and Mats Björkman.

Of the designed products we encounter daily, many have a biotechnology origin although most people do not recognize them as being designed using materials or methods derived from biotechnology. This could concern products such as fermented food and beverage, biological drugs, or diagnostic tests used by the medical care unit in the aftermath of flue.

These biotechnology products are examples of design that includes a wide range of considerations, of course, not only from biology but also from physics and chemistry.

All modern design is, with few exceptions, based on scientific laws and principles. Previous experiences are almost always considered when a new product is designed. Sometimes, a product design can be based on a new invention or a discovery. This is an aspect that is often present for biotechnical products. Biotechnical innovations are, in comparison with many other fields, often based on new inventions or scientific discoveries. This puts extra demands and strains on the design process, as the design task is more complex and complicated compared to the design of a new product that is based on an existing product or range of products. In the latter case, there exist more experience and knowledge of the utilized technology. Furthermore, the scientific basis of the product is better known.

The design concept is not limited to physical artifacts or devices – design is also required for manufacturing processes to produce the designed products and services to support them. Biotechnology products are examples of that.

Design in industry has in recent years gone through a significant development and vitalization in order to strengthen product development in a world of increasing global competition, higher demands from customers, and with tighter regulations. One example of this development is the trend among the producing companies of selling products not only as a single physical unit but also in combination with a service related to the physical product [1]. Furthermore, this trend goes in the direction that the relative value of the service part increases while the physical product part decreases. However, the proportion between the value of the physical product and the service differs very much between different product categories. Traditionally, the service part has often been added to the physical product offer to the customers.

The trend has led to an outspoken industrial interest augmented by significant research efforts on how to integrate the design with a combination of the physical product and its required service. This is often referred to as Product Service Systems design [2]. This new approach means new challenges for the producing companies [3] such as designing their physical products to fit Product Service Systems. In addition, the Product Service Systems approach gives opportunity to close the material flows with product remanufacturing in an economic and environmentally beneficial manner [4]. One of the large business incentives for the producing companies is that the customer relationship is improved that also increases the possibilities for product remanufacturing within the product life cycle [5].

Another aspect of design that grows in importance is the sustainability of the designed product. Sustainable design aims to design products that support a *sustainable development* in society. In 1987, the World Commission on Environment and Development (WCED) defined sustainable development as "a development that meets the needs of the present without compromising the ability of future generations to meet their own need" [6]. As a consequence, sustainable design has three dimensions that must be addressed. The product must be environmentally, economically, and socially sustainable.

The examples in this book address all these dimensions as biotechnology products are intended to improve the health and well-being of people. The sustainability aspect is explicitly not discussed, yet it has a strong impact through economical use of biomechatronic products.

1.2 DEFINITION OF BIOMECHATRONIC PRODUCTS

We refer frequently in this book to *mechatronics*. A mechatronic product is a product where the fields of mechanical, electronic, computer, control, and systems design engineering are combined in order to design a useful product [7,8]. Most of our more advanced consumer and business-to-business products are mechatronic products comprised of combinations of mechanical and electronic components [9]. We may think of a car as a mechanical product. That was true for a Ford Model-T, but today a modern car is a highly advanced mechatronic product where the value and cost of the mechanical components are continuously decreasing in relation to the electronic components and subsystems. Many of our "mechanical" consumer products are in fact controlled by microcomputers/microprocessors. A relatively inexpensive product such as the modern digital consumer camera is based on highly advanced microchip technology for creation of a digital image and the camera is controlled by advanced electronics and microprocessors. Digital cameras are intended for a mass consumer market and the advanced key components are mass-produced. The result is that the development and manufacturing costs can be distributed over a vast number of individual cameras decreasing the unit price of the cameras that, in fact, are highly advanced mechatronic products. *Cheap* does not have to imply *simple* anymore.

To combine and synthesize expertise from the fields of mechanical, electronic, computer, control, and systems design engineering in a design process in order to design a product is not an easy achievement. The major task is to come up with an optimal combination of the different fields of engineering and technologies. This is accompanied with a huge risk for suboptimizations.

A biomechatronic product, as this book focuses on, is a mechatronic product with a substantial element of biotechnology added that shares all these characteristics of the mechatronic product.

In order to emphasize this, we define the biomechatronic products in the following way:

> A product where biological and biochemical, technical, human, management and goal, and information systems are combined and integrated in order to solve a mission that fulfils a human need. A biomechatronic product includes a *bio*logical, a *mecha*nical, and an elec*tronic* part.

The generic biomechatronic definition used here pinpoints a unique feature in design that a biological system in the product is an active part of the design concept. By that, our definition of a biomechatronic product focuses on what systems are included and constitute important parts of the product [10].

Common biomechatronic products include, for example, many products used for production of food and medicine, medical analysis equipment, and so on. This book focuses on these types of biomechatronic products.

1.3 PRINCIPLES OF BIOMECHATRONICS

Figures 1.1 and 1.2 represent the biomechatronic system. A biomechatronic product can be seen as a physical realization of a biomechatronic system. The biological part exerts activity on the mechanical and electronic parts. The mechanical part exerts activity on the biological part and/or the electronic part. The electronic part exerts activity on the mechanical part and/or the biological part. Importantly, external stimuli elicit effects. This may be, as in the figures, directly on the biological part that creates activity or changes activity owing to the stimulus, or it is mediated through the mechanical and/or electronic parts. The tripartite system is required to create the reading (or response).

The abstraction of the object can be applied to a number of today's technical systems in biotechnology.

A biosensor is an example of a biomechatronic system – an analytical device that measures a substance interacting with a biological part, for example, an antibody. Such an interaction may cause a change in activity, for example, a mass change, which causes a change in mechanical oscillation of a quartz crystal, which in turn is electronically recorded by a pair of electrodes. This recording is then transduced to a display reader. The biosensor

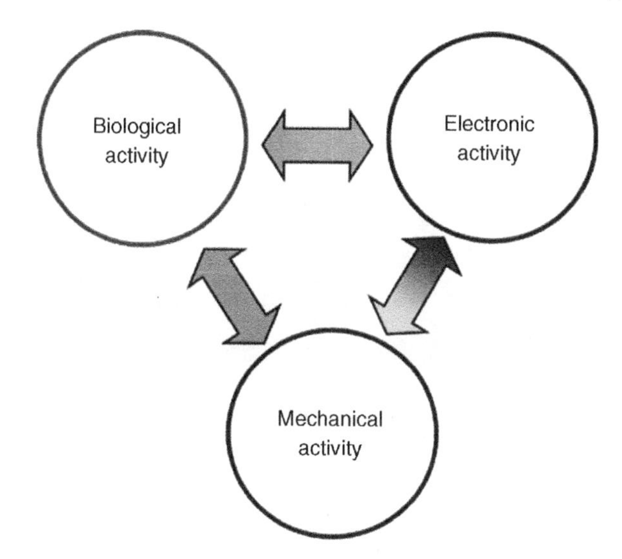

Figure 1.1 *In a biomechatronic system, biological, electronic, and mechanical activities interact in the overall function of the designed object.*

is a so-called quartz crystal microbalance (QCM) as shown in Figure 1.3 where a piezoelectric crystal is oscillating with a frequency determined by the mass load on the crystal [11,12]. The surface of the crystal is covered with a bipolar layer of amphiphilic molecules that mimics biological membrane. For example, nucleotide sequences are immobilized to the membrane and

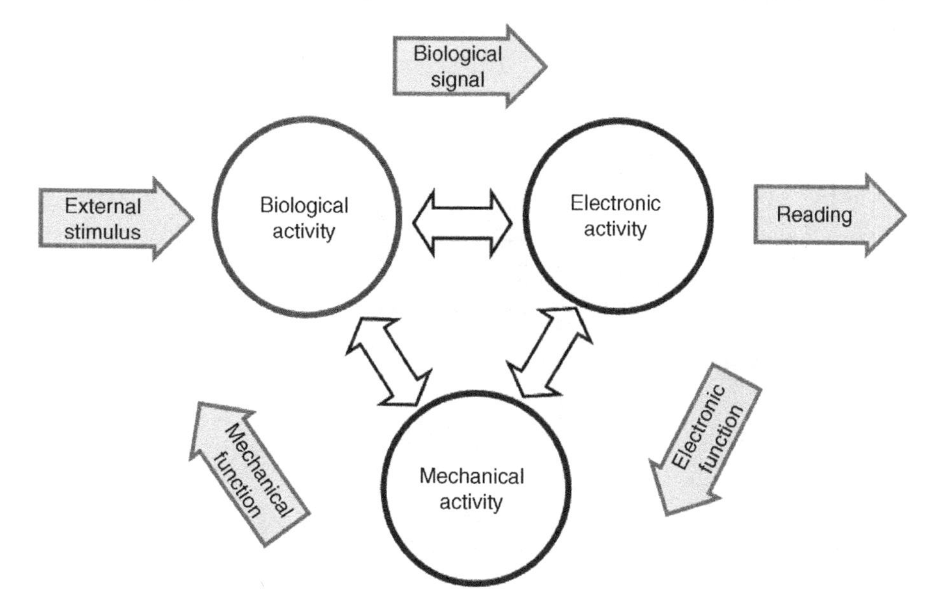

Figure 1.2 *A biomechatronic system in operation.*

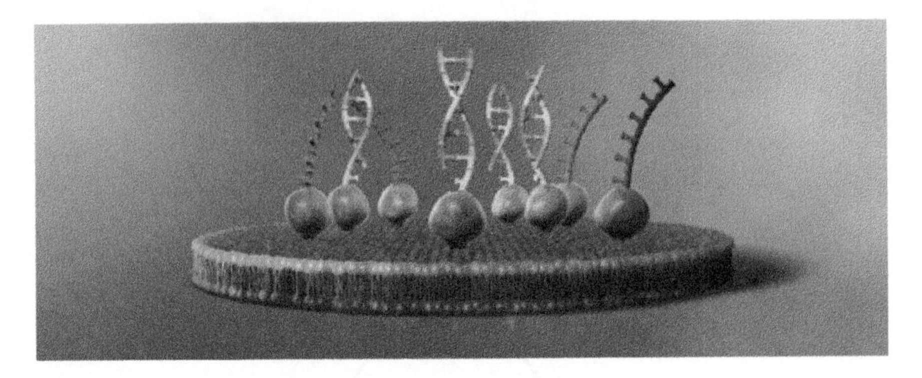

Figure 1.3 A biomechatronic system example: a quartz crystal microbalance biosensor. Reproduced with permission from Q-Sense.

can bind, or hybridize, with complementary DNA sequences. Thus, the QCM is a biomechatronic product since it exemplifies exactly what Figure 1.2 illustrates.

A bioartificial liver [13] is another example of a biomechatronic system – a contained liver cell culture converts xenobiotics/drugs by its metabolic activity (Figure 1.4). A mechanical system, consisting of pumps and valves, directs nutrients and patient blood to the liver cells entrapped in a plastic cage that allows liquid perfusion. Electronic circuits measure activity and control pumps and valves and by that the liver. Human liver cells are cultured on a tubing network of gas- and liquid-permeable channels with a structure mimicking the liver tissue. The system has, as shown in Figure 1.4,

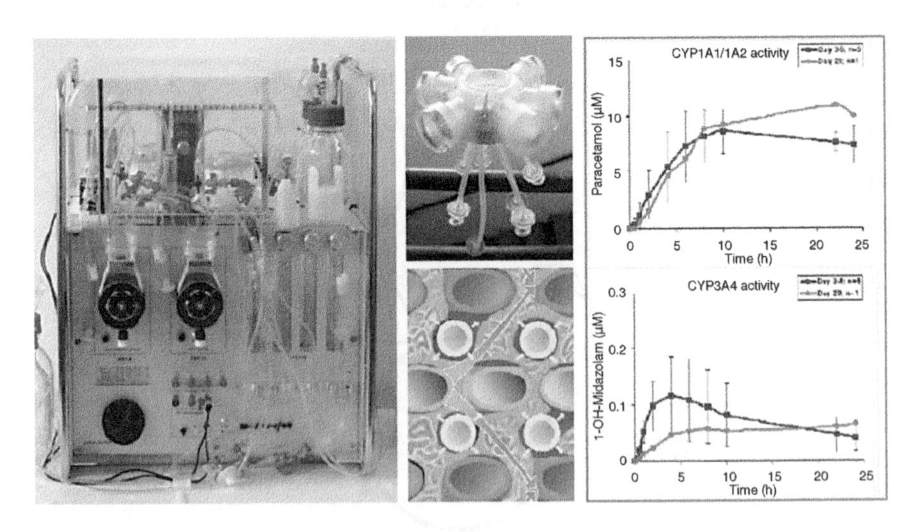

Figure 1.4 A biomechatronic system of a bioartificial liver device [8]. Reproduced with permission from Charité Universitätsmedizin, Berlin.

a mechanical part consisting of pumps, valves, and plastic containers and an electronic part consisting of sensors and actuators that interact with the biological liver cells growing and metabolizing nutrients perfused through the system. The tripartite system is fully interdependent and, by that, also adheres to Figure 1.2.

1.4 BRIEF HISTORY OF THE DEVELOPMENT OF BIOMECHATRONIC PRODUCTS AND ENGINEERING

From a historical perspective, biomechanical products and systems have been created in societies very early. The electronic part of biomechatronic products, however, was developed first during the twentieth century after the technical development of electricity at the end of the nineteenth century.

Biology and mechanics were merged already at least 4000 years BC [14]. Artifacts and excavations from ancient Mesopotamia and Egypt show examples of breweries and wineries on an industrial scale. Recipes for production have been interpreted from hieroglyph tables and cave paintings show snapshots of equipment. This was actually a sort of early Standard Operation Procedures (SOPs) [15]. During the ancient Roman period, factories were set up for industrial olive oil and vinegar manufacture with hundreds of workers.

These historical examples are, in a wide sense, biomechanical inventions. They brought in the grape culture at an industrial level and combined it with technical mechanical devices. Of course, electronics was not applied at this developmental stage.

Also in the field of practical medicine, the biology and mechanics were integrated. In ancient Egypt, medicine reached sometimes a surprising high technical level – in some respects, the medical treatments were technical although the biological understanding was shallow.

Another early example of a sort of bioreactor is the beehive. The biological component – the bee – carried out the transformation from nectar to honey. The technical invention, the hive, nourished the bee swarm and recovered the product.

Brewing and fermentation technology developed slowly until the nineteenth century when Louis Pasteur finally was able to bring a scientific understanding to biological transformation on a molecular and cellular level [14]. Early twentieth-century microbiology refined this knowledge for new microbial strains and products. Electricity became a way to transform energy in the biological system that previously was limited to mechanical functions. Thus, temperature control of a culture could be done and mixing devices could be integrated into the first generation of bioreactors 100 years ago.

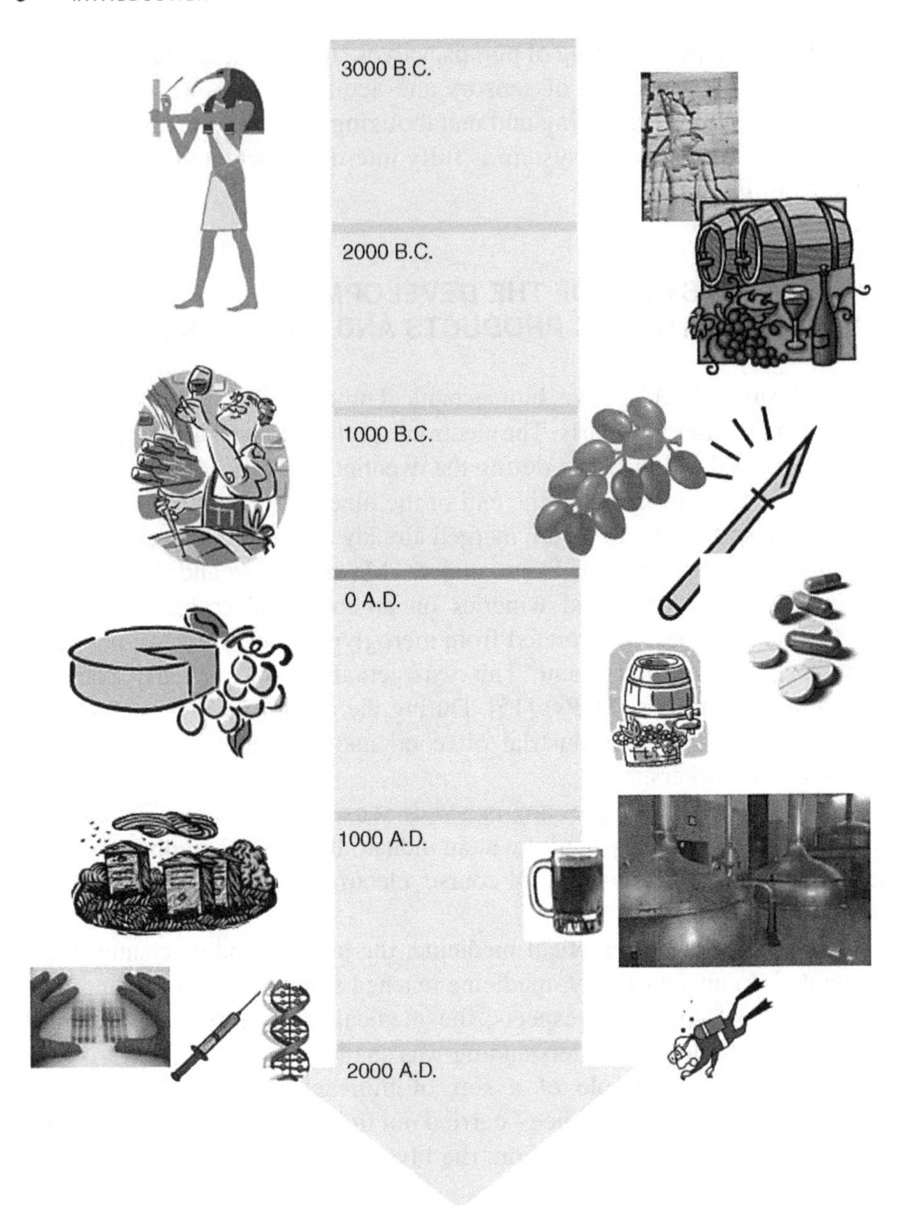

Figure 1.5 *The historical development of biomechatronic products and production systems.*

The timeline of Figure 1.5 illustrates the connectivity of the biotechnology and history. What could be added is that the progress in biotechnology development seen in this timeline was the result of scientific discoveries done in parallel [16,17]. These achievements were not possible without the development of natural sciences and can be associated with names such as

Linnaeus, Jenner, Mendel, Leeuwenhoek, Darwin, Watson, and many others [17–19].

Its seems very likely that the continuation of the time arrow of Figure 1.5 will in the near future be decorated with numerous new inventions and follow-up products extruding into the micro- and nanoscale world of biotechnology. If this development is augmented by systematic design approaches, the new biotechnology products have a better potential to gain high quality at an earlier stage of development.

1.5 AIM OF THIS BOOK

The overall aim of this book is to support the task of designing biomechatronic products. It is very complex to design mechatronic products and the task is further increased in complexity when the biotechnological element is added in the form of biological activities and systems. This is illustrated by the bioartificial liver above. The focus in the book is on how to integrate and handle this biotechnological element in the integrated design process for the design of biomechatronic products and systems.

We try to fulfill the aim by several means. As the major enabling means, we propose and present a generic methodology for systematic design of biomechatronic products. This biomechatronic design methodology includes a number of design tools for supporting the work in the different stages or parts of the methodology. This is mainly done in Chapter 4, which can be seen as the central chapter of this book.

The presented methodology and design tools are all based on existing and well-established methodologies and design tools from the field of mechanical engineering. These are presented mainly in Chapters 2 and 4. The methodologies and tools are further developed and adapted to include the biotechnological elements of biomechatronic design. Biotechnology in mechatronic design is treated mainly in Chapters 3 and 4.

One important aspect of the presented biomechatronic design methodology is that it gives the means for supporting the communication between experts in the different fields of engineering that are involved in the design. Many suboptimizations in designs are the result of misunderstandings in the communications.

The presented methodology and design tools are applied to a number of biomechatronic products and/or systems. This is covered in Chapters 5–13. These applications give an insight and understanding of how the methodology could be applied and used.

The applications in Chapters 5–13 also give comprehensive descriptions of many different types of modern biomechatronic products and systems. These descriptions can be of great interest, especially for readers unfamiliar with biotechnology products, regardless of the applications of the biomechatronic methodology.

It is possible for the reader to focus on the chapters that describe biomechatronic products and systems of special interest for the reader. However, we strongly advise the reader to also read the other chapters, as all applications are not implemented in the same manner. The different applications act as examples of how it is possible to apply and utilize the presented methodology and design tools in different ways. There is not just one way of using them. This is clearly illustrated in the different application chapters.

REFERENCES

1. Lifset, R. (2000) Moving from products to services. *J. Ind. Ecol.* 4, 1–2.
2. Sundin, E., Lindahl, M., Ijomah, W. (2009) Product design for product/service systems: design experiences from Swedish industry. *J. Manuf. Technol. Manage.* 20, 723–753.
3. Sundin, E., Ölundh Sandström, G., Lindahl, M., Öhrwall Rönnbäck, A., Sakao, T., Larsson, T. (2009) Challenges for industrial product/service systems: experiences from a learning network of large companies. In: *Proceedings of CIRP Industrial Product/Service Systems (IPS2)*, Cranfield, UK, April 1–2 pp. 298–304.
4. Sundin, E., Bras, B. (2005) Making functional sales environmentally and economically beneficial through product remanufacturing. *J. Cleaner Prod.* 13, 913–925.
5. Östlin, J., Sundin, E., Björkman, M. (2009) Product life-cycle implications for remanufacturing strategies. *J. Cleaner Prod.* 17, 999–1009.
6. United Nations (1987) Report of the World Commission on Environment and Development. General Assembly Resolution 42/187, December 11.
7. Bradley, D.A., Loader, A.J., Burd, N.C., Dawson, D. (1991) *Mechatronics: Electronics in Products and Processes*. Chapman & Hall, London.
8. Karnopp, D.C., Margolis, D.L., Rosenberg, R.C. (2006) *System Dynamics: Modeling and Simulation of Mechatronic Systems*, 4th edition. Wiley.
9. Cetinkunt, S. (2007) *Mechatronics*. Wiley.
10. Derelöv, M., Detterfelt, J., Björkman, M., Mandenius, C.F. (2008) Engineering design methodology for bio-mechatronic products. *Biotechnol. Prog.* 24, 232–244.
11. Ngeh-Ngwainbi, J., Suleiman, A.A., Guilbault, G.G. (1990) Piezoelectric crystal biosensors. *Biosens. Bioelectron.* 5, 13–26.
12. Fung. Y.S., Wong, Y.Y. (2001) Self-assembled monolayers as the coating in a quartz piezoelectric crystal immunosensor to detect *Salmonella* in aqueous solution. *Anal. Chem.* 73, 5302–5309.

13. Gerlach, J.C., Lübberstedt, M., Edsbagge, J., Ring, A., Hout, M., Baun, M., Rossberg, I., Knöspel, F., Peters, G., Eckert, K., Wulf-Goldenberg, A., Björquist, P., Stachelscheid, H., Urbaniak, T., Schatten, G., Miki, T., Schmelzer, E., Zeilinger, K. (2010) Interwoven four-compartment capillary membrane technology for three-dimensional perfusion with decentralized mass exchange to scale up embryonic stem cell culture. *Cells Tissues Organs* 192, 39–49.
14. Bud, R. (1993) *The Uses of Life: A History of Biotechnology*. Cambridge University Press.
15. Gaudillière, J.P. (2009) New wine in old bottles? The biotechnology problem in the history of molecular biology. *Stud. Hist. Philos. Biol. Biomed. Sci.* 40, 20–28.
16. Mason, S.F. (1962) *A History of the Sciences*. MacMillan, New York.
17. Prigogine, I., Stengers, I. (1984) *Order Out of Chaos*. Bantam Books, New York.
18. Schneer, C.J. (1960) *The Search of Order*. Harper and Brothers, New York.
19. Watson, J.D. (1968) *The Double Helix: A Personal Account of the Discovery of the Structure of DNA*. Atheneum, New York.

Part I

Fundamentals

2

Conceptual Design Theory

This chapter summarizes fundamental principles of mechatronic design with the mechanical engineering perspective. We present here the concepts in mechatronics upon which are based the biotechnology applications discussed in this book. The chapter does not bring up biotechnology aspects *per se* – this will be saved for the following chapters.

2.1 SYSTEMATIC DESIGN

2.1.1 Design for Products

"The main task of engineers is to apply their scientific and engineering knowledge to the solution of technical problems, and then to optimize those solutions within the requirements and constraints set by materials, technological, economic, legal, environmental and human-related considerations," say Gerald Pahl and Wolfgang Beitz [1], two German researchers who have made significant impact on research, teaching, and training in mechanical design engineering for several decades.

Biomechatronic Design in Biotechnology: A Methodology for Development of Biotechnological Products, First Edition. Carl-Fredrik Mandenius and Mats Björkman.

However, the scope of design is not restricted to mechanical engineering, it applies to all engineering sciences and its basic principles are as much applicable to engineering design problems and solutions in electronics, physics, information technology, chemical technology, and biotechnology, for example.

Technical problems become concrete engineering tasks after their clarification and definition that engineers have to solve to create new technical products. These products may vary widely: mechanical products, electric devices, power plants, chemical processes, software programs, and so on.

The steps in the creation of the products are several, for example, identification of needs and constraints, the original idea, the conceptual elaboration of the idea, considerations of alternative ways to realize the idea, the building or construction work of a first prototype, and the (mass) manufacturing of it (Figure 2.1).

The invention of a new product is the typical main achievement of design and development engineers, whereas its physical realization is the main accomplishment of manufacturing engineers. However, it is very important that there is a close cooperation between the development engineers and the manufacturing engineers during the product realization process. The design of the product and the design of its manufacturing system must be performed concurrently in an integrated process. This is vital in order to increase the possibility of achieving not only a product that is adapted and suitable for manufacturing but also an efficient manufacturing process and system (Figure 2.2).

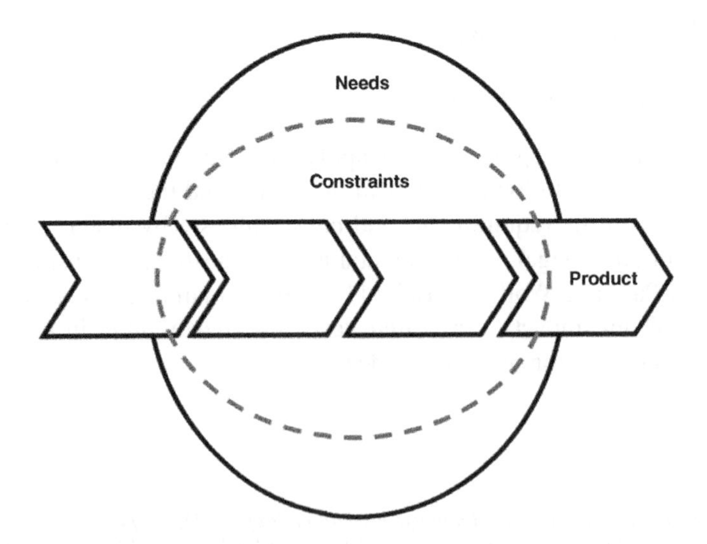

Figure 2.1 *Principle for stepwise product design under needs and constraints.*

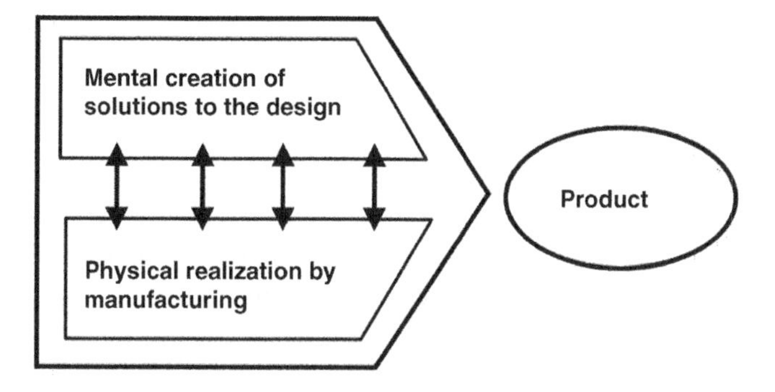

Figure 2.2 *The concurrent development of design solutions and manufacturing process.*

It is clear that design of products is an engineering activity that

- has the potential to affect almost all areas of human life;
- should take advantage of using the laws and insights of science;
- can be built upon specific experience;
- provides the prerequisites for the physical realization of novel ideas.

The approach in this book is that design should be done systematically. This is the dominating approach in both industry and academia. Another way to say this is that in *systematic* respect, designing is the optimization of given objectives within partly conflicting constraints [2]. User requirements changes gradually, so a particular solution can be optimized only for the actual circumstances.

Nowadays, it is considered important to design new products with an eye on the whole *product life cycle* (Figure 2.3). The sustainability aspect, mentioned in Chapter 1, is an example of this. This introduces an extended view to designing – in such organizational respects the design can be an essential part of the product life cycle. This cycle is triggered normally by either a market need or a new idea, or both. It starts with product planning and ends when the

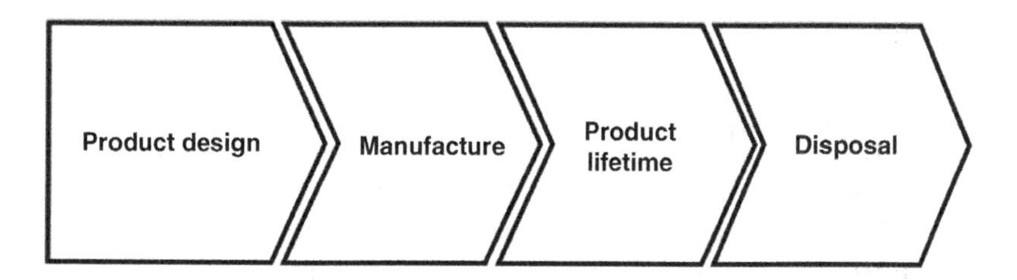

Figure 2.3 *The product life cycle.*

product's useful life is over, with recycling/remanufacturing or environmentally safe disposal.

2.1.2 Origin of the Design Tasks

It is common that systematic design projects, related to (mass) production or batch production, are initiated by a product planning team after carrying out a thorough analysis of the market needs. A number of issues are essential to consider:

- *The Novelty of the Product:* Does a similar product already exist? Biotechnology has in this respect been favored for entering a virgin market area with many opportunities. The product may be based on a new scientific discovery. This is not uncommon for the biomechatronic products. The novelty aspect has therefore often been easy to accomplish. With a maturing biotechnology market segment, this has become more difficult.
- *The Production Cost:* It must be realistic and within the tolerance limit of the market.
- *The Complexity of the Product:* The design applications in coming chapters address the complexity of biomechatronic design. However, biomechatronic products are not the sole complex product type – large-scale mechanical systems or mechatronic products can often be very complex products themselves before merging with the biotechnical dimension.
- *Realistic Goals:* Goals must be realistic and must not cross the limits of aspirations. This is both a restriction and a challenge for the creativity and imagination of the design team. Some of the products we will discuss approach this limit, such as artificial organs, stem cells, and biological microchips.

The essential considerations above create a number of working tasks and activities for the design team that should be approached systematically. But first, let us bring up a few historically established design issues that have driven design achievements over the past years and shaped the way we apply design today.

2.1.3 Development of Design Thinking

From a recent historical perspective, it is possible to mention a few steps of particular importance for the ascent of design thinking.

An important theoretical contribution was the identification of governing scientific principles and economical constraints that conduct systematic

design behavior [3]. Concerning the form of the design, these principles are related to (1) *minimizing production cost* in order to improve the competitive edge of the design solution, (2) *minimum space*, which is close to the former, where it is assumed that the smallest product that can fill the same need is the most attractive solution, (3) *minimum weight*, another criterion that strengthens the product is that it is reasonably inexpensive, (4) *minimum losses* in the production of the product, and (5) *optimal handling* when using the product.
In the design and optimization of individual parts and simple technical artifacts of the product, these principles could and should be followed.

Another valuable contribution was the introduction of the four design perspectives that are applicable to most new products and consequently are the fundamental design factors (Figure 2.4). These are (1) working principle, (2) material, (3) form, and (4) size of the design object [4]. All designed products can be described from these aspects. They are interconnected and dependent on needs, production volume, production costs, and a couple of other constraints. The four aspects are by necessity investigated and developed in this sequence. First, identify and decide on the working principle, then select suitable materials, and finally consider the form and the size.

Another approach to systematizing the design was to start with an overall layout and then set main dimensions and general arrangements [5]. Then, this overall design was divided into parts that could be handled in parallel in the

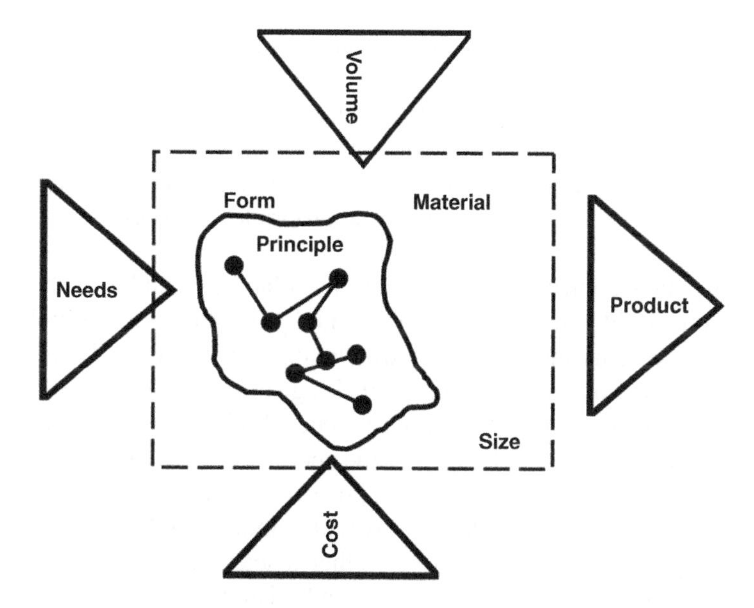

Figure 2.4 *The four design perspectives suggested by Tschochner.*

design work process. When doing this, the identified design tasks were systematically varied and with this a number of alternatives could be generated from which the optimal solution could be identified.

Later on, the design working process was divided into distinguishable phases [6]. The first is the establishment of the working principle based on an idea or invention. The second is the embodiment of the idea where the form is laid out and supported by calculations based on scientific rules and knowledge. The third phase is the implementation of the first and second phases. Although seemingly straightforward, these phases were and are occasionally overruled resulting in unfavorable consequences.

An important step in the development of the existing systematic design theory, now often referred to as the *basic systems approach*, was taken in the 1960s [7]. This is based on a thorough and critical analysis of the task (goals, specification, and overall function), followed by a systematic search for solution elements and their combination into a working principle. Subsequently, these combinations are analyzed critically in order to identify shortcomings and find a remedy for alleviating these shortcomings, and by that, present a new solution. This solution is further analyzed, recombined, and optimized. The way to develop a new product has many traits that are further developed in the biomechatronic approach we apply in this book.

2.1.4 Recent Methods

No doubt, several of these early theoretical ascents have been integrated into today's design methodology.

A relatively novel approach that is particularly relevant to biological problem formulation is the introduction of a reasoning based on flows, energy, and signals [8]. Although originally it was concerned with characterizing mechanical systems, it is particularly appropriate to apply on biological systems.

An important contribution was made by Vladimir Hubka in the 1970s [9–11] to the establishment of the fundamental principles of a comprehensive design science. This included the introduction of a common design terminology and the use of common symbols in design diagrams (Figure 2.5). It also introduced a clearer notion of the design process both on an abstract level and in actual tasks. By that, guidelines for the activities of designers were set up and could be applied in industrial design practice. We have chosen to follow up on many of the Hubka principles by involving modules and tools, such as the biological systems entity (as will be further discussed in Chapters 3 and 4 where scientific design terminology and symbolism introduced by Hubka are explained in more detail).

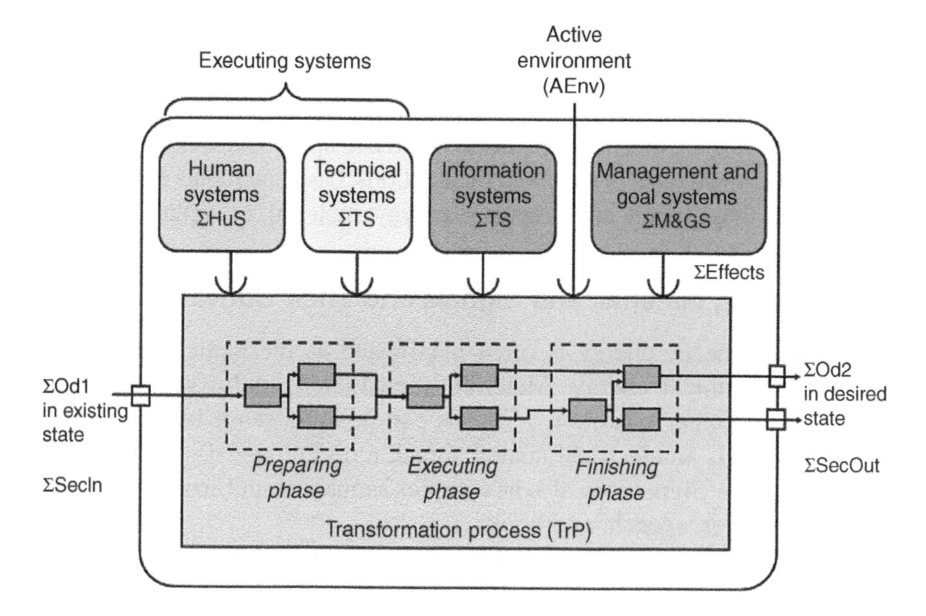

Figure 2.5 *The scientific design terminology and symbolism introduced by Hubka.*

Economical assessment in systematic design is a key to success of the product especially in later stages of the product life. Attempts to improve the possibilities of assessing the cost for design alternative early in the design process have been facilitated by using computer-based calculations and support by databases [12]. By systematically breaking down the parts of the solutions, the necessary details can be reached.

Several other valuable contributions to systematic design have been made during the recent period. Most of them are based on the fundamentals mentioned above [13–16]. Also, these works have inspired the methodology applied in this book.

2.2 BASICS OF TECHNICAL SYSTEMS

The term *technical systems* comprises a wide range of artifacts or collection of artifacts. It encompasses many sciences and technologies, for example, physics and derived branches, including mechanics, thermodynamics, materials science, and so on. It can also include chemistry with chemical technology and reaction engineering, biotechnology and computer science and engineering, just to mention some of them.

All technical tasks are performed by technical artifacts or objects, following certain instructions, executed by a human operator, electronic actuators, or software programs.

A classification of technical artifacts has been suggested by Hubka [10]. This is based on function, working means, complexity, production, product, and other critical conditions. This type of classification is not always useful due to its disparity and varying origin. More useful is often to apply *system boundaries* in which the technical objects will exert their actions, based on inputs and outputs. The form of these inputs and outputs is energy, material, or signals.

2.2.1 Energy, Material, and Signals and Their Conversion

In technical objects, energy is often manifested as mechanical, electrical, optical, or chemical energy. Material is characterized by weight, color, substance, and other conditions. Signals can in some sense be regarded as information; it is an electrical analog signal, a digital signal, or a chemical signal substance. Signals can also be complex sequences and arrays of signals, such as messages, speech, and books.

Energy can be converted from one form to another in the technical system. In a combustion engine, thermodynamic energy is converted to mechanical energy; in a muscle fiber, biochemical energy is converted to mechanical energy; and in a water power plant, hydrodynamic energy is converted to electrical energy.

Materials can be converted as well. Metals can be melted and mixed to produce alloys. Food nutrients can be converted to vital energy in a growing body. Polymers can be shaped to desired forms with specific functions.

Signals can also be converted, or as more commonly said, be transmitted, displayed, recorded, and received. This conversion occurs, for example, in a biological cell in the brain and in the eye when reading a message, in an electrical circuit in a computer, and so on.

Important for design work is that energy takes the form of mechanical, thermal, electrical, chemical, optical, nuclear, force, heat, current, and *biological* energy.

Materials appear in the form of gas, liquid, and solid dust, and also as raw materials, as test sample, and as workpieces. Furthermore, materials also appear as end products and components.

Signals are in design work accompanied with characteristic features, such as signal magnitude, display, control impulse, data, or information. Conversion of energy, material, and signals can normally be related to quantity and quality (Figure 2.6).

2.2.2 Interrelationships of Functions

When describing and solving a design problem, it is appropriate to apply the term *function* to the general input–output relationship of a system that executes a task.

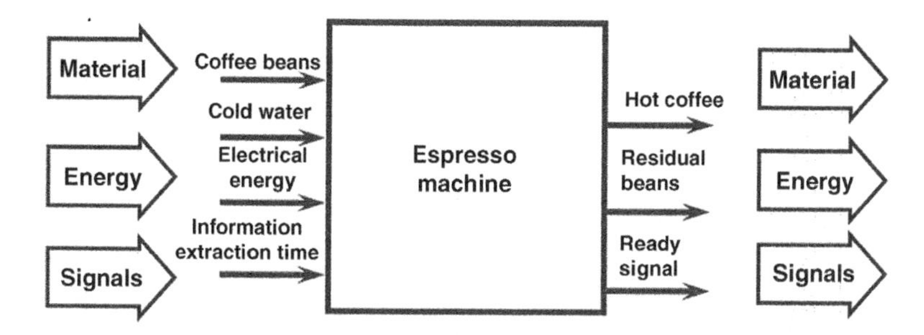

Figure 2.6 *Conversion of energy, material, and signals in a design object.*

This has the advantage of providing a system description with a clear and easily reproduced relationship between these inputs and outputs. Applying this will be very helpful for solving a technical problem.

Once the overall task is well defined, which should be made clear from the inputs and outputs, the overall function of the system can be described.

The overall function can be divided into subfunctions that also have defined subtasks.

In the design work, it is very useful to carefully consider the meaningful and compatible combination of subfunctions into an overall function. This provides the designer with a *function structure*. This structure may be varied to satisfy the overall function (Figure 2.7). The figure illustrates this where it is shown how the structure includes energy, material, and signals.

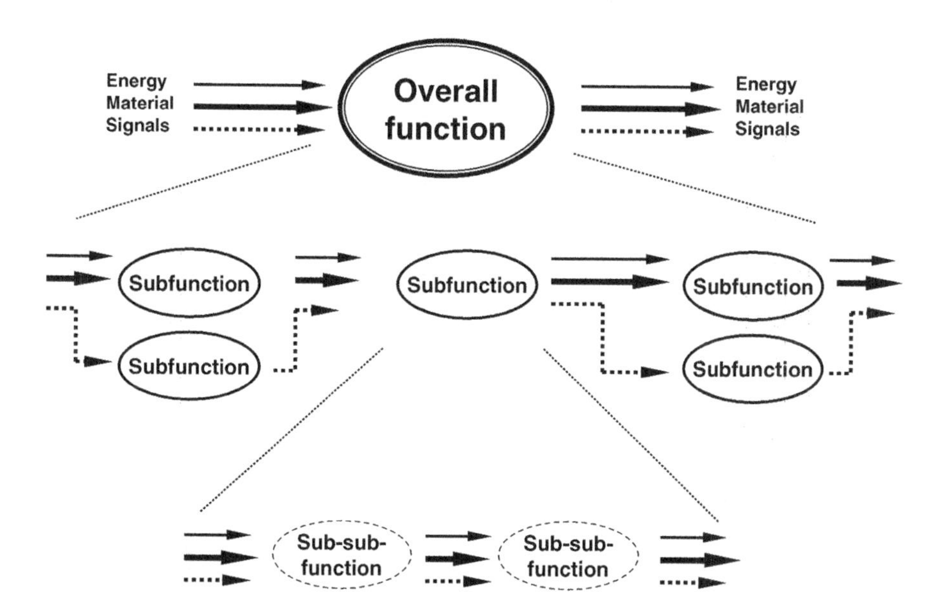

Figure 2.7 *The function structure where the overall function is broken down into a variety of subfunctions that all convert energy, material, and signals.*

TABLE 2.1 Different Levels of Interaction and Interrelationships in an Espresso Machine

Interrelationship	Elements	Structure	Example
Functional interrelationship	Functions	Function structure	Espresso machine Regulate the heating Heat the water Unite tea and water Regulate the infusion Allow the infusion Separate tea and leaves Tea Leaves Transform energy to heat Activate the separation Support subsystems Energy (electricity)
Working interrelationship	Physical effects and geometric and material characteristics Working principle	Working structure	Beans Heater Electronic controller Water tank Grinder (Principle drawing)
Constructional interrelationship	Components Joints Assemblies	Construction structure	A typical espresso machine (Construction drawing)
System interrelationship	Artifacts Human beings Environment	System structure	Espresso machine Water supply Bean plantation Drinker Milk Sugar supply CE certification (System drawing)

2.2.3 Interrelationship of Constructions

A concretization of the function structure leads to a construction structure. By this, the subfunctions are embodied by physical, chemical, or biological processes. A majority of subfunctions are of mechanical or electrical/electronic engineering nature for most products. Thus, objects where mechanical and electrical parts are the main design issues dominate over the others. So far, biological processes are almost absent in the theoretical design literature.

What is well studied and described is the physical process. This is realized by physical effects occurring under certain geometrical and material conditions. This results in a working interrelationship that fulfils the function necessary for performing the task.

A further concretization of the working interrelationship paves the way for a construction structure of the design. For this interrelationship, the modules, assemblies, and machines with their connections deliver a more concrete technical system.

2.2.4 Interrelationship of Systems

Since the technical system normally is a part of a large technical product context, a design cannot be successful without considering this higher level. This will include human beings, laws of the society, and the surrounding physical environment. In systematic design, this level is referred to as the system interrelationship. It is especially useful for studying effects from the interacting environment, humans, unpredictable effects, external information, and so on. Table 2.1 shows the levels of interactions we have discussed illustrated with an example from mechatronic engineering – an espresso machine. In fact, this also has a biological component in its input material.

2.3 PSYCHOLOGY IN THE SYSTEMATIC APPROACH

The systematic design approach has its roots in the human psychology and our inherited way of thinking.

Often, we form our thinking along the relationship of concrete with abstract descriptions. The way to move our ideas from one concrete solution to another concrete solution goes over a state of abstraction.

Another relationship of importance is the whole and the parts. We have the ability to dissect the wholeness of an idea into its parts, but we also have the ability to bring together parts to the wholeness, or even to another wholeness than we originally dissected (Figure 2.8).

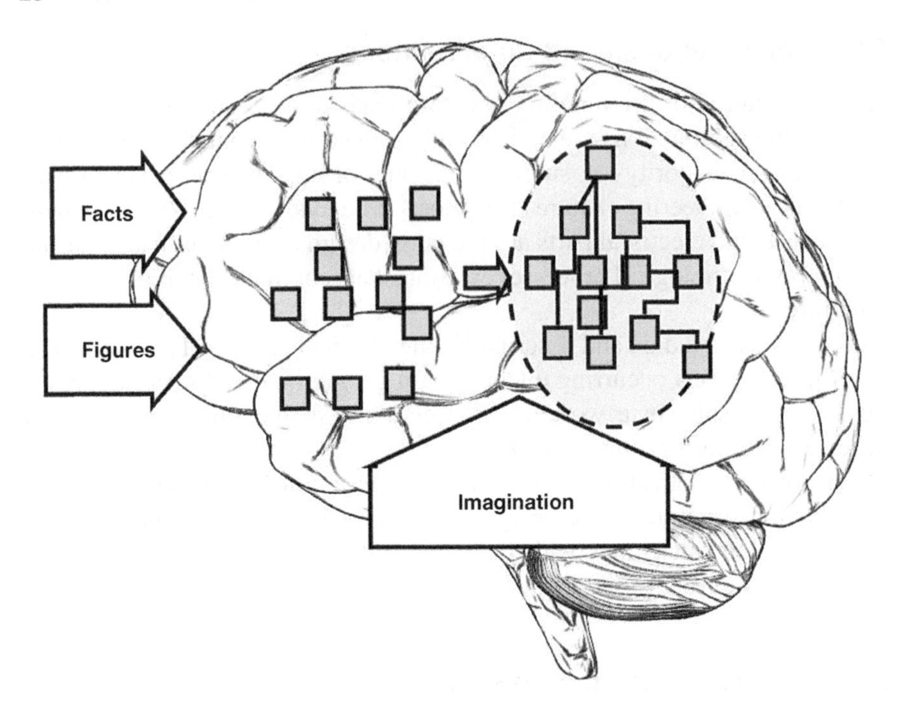

Figure 2.8 The process in the brain during systematic design.

A third relationship that influences design thinking is space and time. In what order should the parts operate and where in space?

These relationships can be investigated more easily with the help of a systematic approach.

Intuitive thinking is an element of designing. Intuition is strongly associated with flashes and inspiration. It happens to a large extent unconsciously. Insights appear suddenly, sometimes after unconscious thinking. The result of the intuitive thinking requires to be systematically investigated.

Discursive thinking, the conscious process that can be communicated and influenced, is easier to structure. Facts and relationships are consciously analyzed, varied, combined in new ways, checked, rejected, and considered further.

A combination of the above is very beneficial but not so easy to achieve during the design process.

2.4 A GENERAL WORKING METHODOLOGY

A systematic approach strives to present a methodology that on a general level is working well in most applications [2]. It should be applicable to a wide range of situations. Actually, not only classical engineering products are under consideration but also services, organizational activities, company business systems, and so on.

Is it possible to work using the same general approach for most applications when carrying out the design work? The theory believes that. According to Pahl and Beitz there are six preconditions that must be satisfied for all systematic approaches [1]:

1. Formulating the overall goal and subgoals for the system. The goals become the driver for the work.
2. Clarifying boundary conditions for the work from initial constraints.
3. Dispelling prejudice that risk to limit the possibility to see new solutions and others cause illogical setbacks.
4. Searching variants to the first solution so that there is a multitude of choices to select from and to compare.
5. Evaluating the variants from the previous step thoroughly versus the goal.
6. Making decisions about the evaluated variants based on the goal compliance.

This is not so easy to carry out as the short list may imply. A number of additional criteria and conditions should be met.

As mentioned in Section 2.3, intuition in combination with discursive thinking is a potential resource of new ideas. The limitation is here that intuitive ideas are not possible to generate by organizing actions, it is time independent and requires individual talents. Furthermore, it is difficult to train this ability.

One should also actively try to minimize errors in the design [2]. This could be done by

- careful definition of the requirements and constraints of a task;
- not persisting on intuitive solutions without a combination of discursive evaluation;
- avoiding standard or fixed ideas;
- avoiding going for the easiest solution available at the moment.

Other means are also useful, for example, allowing new ideas to mature, having a clear and realistic time planning for the work, and systematically working along the relationships of abstraction–concretism, time–space, and whole–parts.

2.4.1 Analysis for Resolving Technical Problems

Analysis is a powerful tool for resolving complex tasks. Problem analysis is the identification of the essential bottlenecks for a successful solution.

Structure analysis is the action of bringing order and logical connections into the complex of problems associated with a task. It usually results in a hierarchical structure of the problems. This may better reveal how they are related and how solutions can be found from other systems.

It is also helpful to apply weak spot analysis of the system: Where may it fail and why? Could the weak spots be ranked in terms of severity? And could a remedy be anticipated?

2.4.2 Abstraction of Interrelationships of Systems

Another means is to apply the principle of abstraction on the system. This tends to reduce the complexity and to generalize the problems. Abstraction means that the problems are placed on a generalized abstract level. It simplifies to see parallel solutions from other systems that share the same level of abstraction. This may stimulate creativity and systematic thinking and reduce the risk of going for an easy concrete solution too soon.

2.4.3 Synthesis of the Technical System

By synthesis in designing we mean assembling of the parts that shall create the system. Normally, it is possible to generate by synthesis a number of solutions. A systematic variation is here of value. However, the risk is that too many permutations result. Some are obviously redundant and unrealistic. For the trained engineers, this can be considered trivial. On the other hand, a risk of introducing prejudices exists in the sorting process.

2.5 CONCEPTUAL DESIGN

The term conceptual design refers to the identification of the essential design problems by applying abstraction, function structures, and working principles and their combination into a system. Conceptual design is by that a way to generalize the design problem. It creates a principle for its solution.

The initiating event is a decision that something is going to be designed. From that decision certain questions must be asked. These are as follows:

- Is the design task clear enough?
- Should more information be gathered?
- Is the goal tangible with existing resources?
- Is it needed or does it actually already exist?
- How should the design be done systematically?

Conceptual design can be divided into nine steps after the task has been defined until the conceptual solution is ready [1]. The steps are the following:

1. *Specification:* A specification is established for the design task. This is mainly based on needs of the potential users and customers.
2. *Abstraction in Order to Identify Essential Problems:* The specification is brought to an abstract level in order to simplify the identification of the most essential design problems to be solved during the conceptual design.
3. *Establishing Function Structures:* The overall function structure is established and the subfunctions within the overall functions are identified and established.
4. *Searching Working Principles to Fulfill Subfunctions:* The possible working principles for the subfunctions are considered. A selection of these is identified.
5. *Combining Working Principles into Working Structures:* The possible working principles are considered and a working structure is set up.
6. *Selecting Suitable Combinations:* By using selection tools the possible combinations are investigated, assessed, and ranked.
7. *Generating Principle Solution Variants:* Find new variants of solutions that are different from the first. Use the experience of the previous steps.
8. *Evaluating Variants Against Technical and Economical Criteria:* Take all variants into consideration and assess their strengths toward each other.
9. *Principle Solution (Concept) Presented:* The variant with the best ranking is taken as the principle solution.

The steps and their role in the process are shown in Figure 2.9.

2.6 ABSTRACTION IN ORDER TO IDENTIFY ESSENTIAL PROBLEMS

As mentioned, abstraction is a powerful way to create new solutions. It intends to enforce new conceptual ideas and to negate those that rely too much on traditional solutions and methods. New ideas should solve an essential existing problem. A key issue is to identify the existence of the essential problems. The most essential problems are often not the most obvious ones.

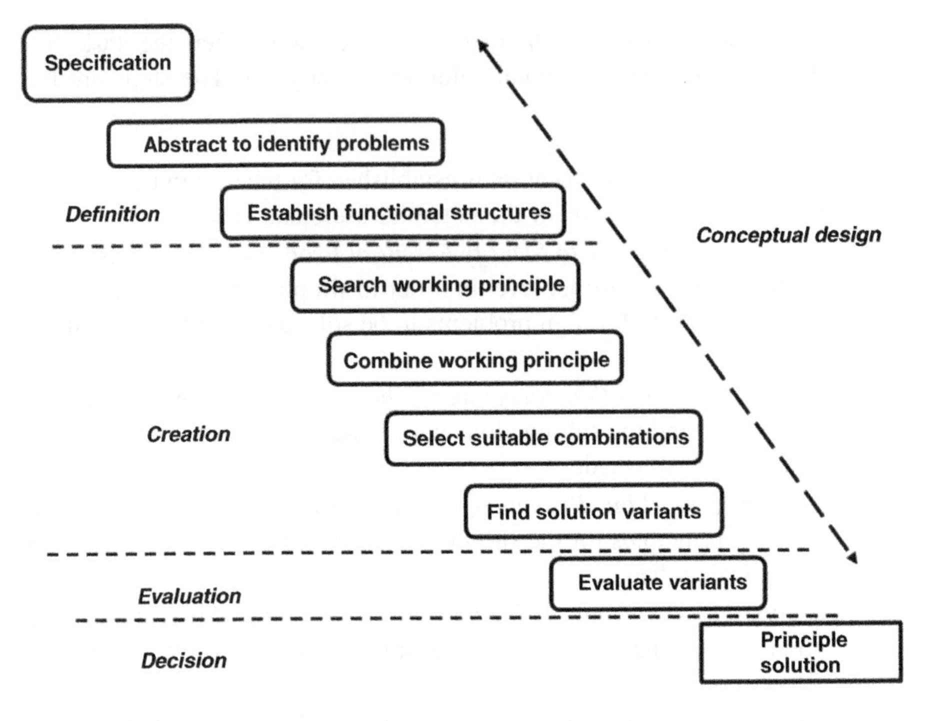

Figure 2.9 *The steps of conceptual design as suggested by Pahl and Beitz [1].*

It is important to analyze the requirement list. That can be analyzed by a stepwise abstraction procedure (Figure 2.10) to reveal the general aspects [1]. These steps are as follows:

1. Eliminate personal design preferences. Be aware of when these appear. Avoid repeating old standard solutions of your own.
2. Omit requirements that have no direct bearing on the function. These can easily take over the function. They play no role in an abstraction of the essential problems.
3. Transform quantitative data into qualitative data of the solution in the form of statements about the system and problem.
4. Generalize the results of these statements to a more fundamental conclusion.
5. Formulate the problem in such a way that it does not influence the choice of solutions in a subjective way.

Once the overall problem has been formulated, it is possible to set up a rough diagram that describes the overall function where the flows of energy,

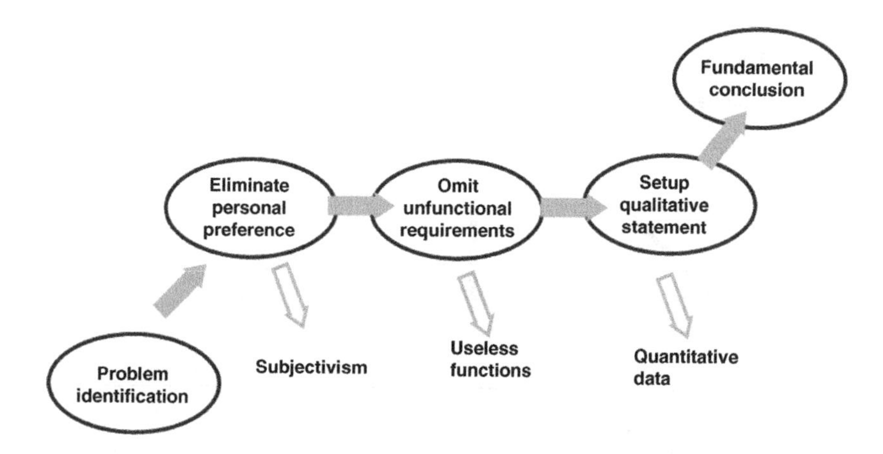

Figure 2.10 *The process of conceptual abstraction.*

material, and signals between the inputs and the outputs of the system are expressed. This structure should be more detailed than the previous one.

Initially, the diagram will inevitably be opaque. When we break it down into subfunctions, the full understanding of the design is gained.

The optimal method of breaking down an overall function – that is, the number of subfunction levels and the number of subfunctions per level – is determined not only by the relative novelty of the problem but also by the method used to search for a solution.

If it is an original design, neither the individual subfunctions nor their relationships are generally known. Then, the search for and establishment of an optimum function structure constitute some of the most important steps of the conceptual design phase. In the case of adaptive designs, on the other hand, the general structure with its assemblies and components is much better known, so that a function structure can be obtained by the analysis of the product to be developed. Depending on the specific demands of the requirements list, that function structure can be modified by the variation, addition, or omission of individual subfunctions or by changes in their combination. Function structures are of great importance in the development of modular systems. For this type of variant design, the physical structure, that is, the assemblies and individual components used as building blocks and also their relationships, must be reflected in the function structure (Figure 2.11).

2.7 DEVELOPING THE CONCEPTS

The conceptual design may suffer from not being concrete enough to generate a definite concept that can be taken further toward production. If only

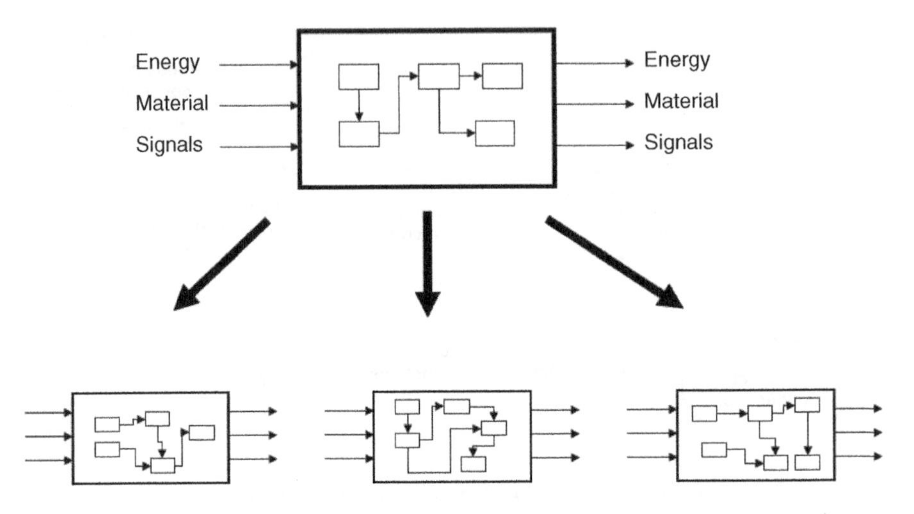

Figure 2.11 Establishing the variant of the functions of the design.

functional structures are evaluated, a gap is created in the blueprint solutions. More concrete qualitative and quantitative definitions must be provided.

This may be attained with rough calculations with assumptions, rough sketches or scale drawings of possible layouts, preliminary experiments or model tests to determine main properties, construction models to aid analysis and visualization, systems simulation, search for patents and literature data, market enquiries of proposed technologies, materials, parts, and so on.

This will generate new variants. Evaluation of these variants from manufacturing and cost perspective will now be easier. The criteria for evaluation should then be carefully established. A list of requirements or needs based on user opinion is a part of this process. Economical characteristics and constraints are other criteria. Target values based on assessed demands become a very valuable complement to this. Methods of weighing the importance of the criteria are other complementary facilities highly important for the outcome of the evaluation.

Procedures for assessing the capacity and quality of conceptual solutions must be solid. These procedures shall allow assessment of a number of parameter values that finally will end up in an overall value that determines where a solution shall be prioritized.

This type of analyses are preferably done with a set of evaluation matrices where the parameters to be evaluated are on the axes of the matrix.

The values become coordinates of the matrix and by that an overview is obtained. The matrices can be expanded into multidimensional systems and include weighing factors. It is certainly open for introducing statistics and other mathematical algorithms in the assessment.

Still, data in and tuning of weights decide the outcome. Thus, proper experiments, enquiries, model evaluations, and so on are crucial for these tools.

2.7.1 Organizing the Development Process

The principles described above have also been used by Ulrich and Eppinger [17] who have established a generic development process. A road map (Figure 2.12) has been suggested that has a number of properties in common with the general theory but is more strongly integrated into the design development work of the designer team. It emphasizes the iterative way of working in product development and it also integrates into the process the manufacturing constraints. A number of tools are provided that contribute to turn the work into a well-planned mechanistic procedure easy to monitor and control.

This methodology was mainly intended for mechanical engineering design [17], which applies to an engineering team at any firm involved in product development. The Ulrich–Eppinger approach presupposes that the team design mission is to establish and manufacture a new product. Success on the market is a key consideration. Thus, the design process starts with the identification of the customers' needs from all aspects: market, users, and manufacturing requirements. The needs are normally verbal and qualitative. The next step is to specify the needs in detail and try to apply metrics to them.

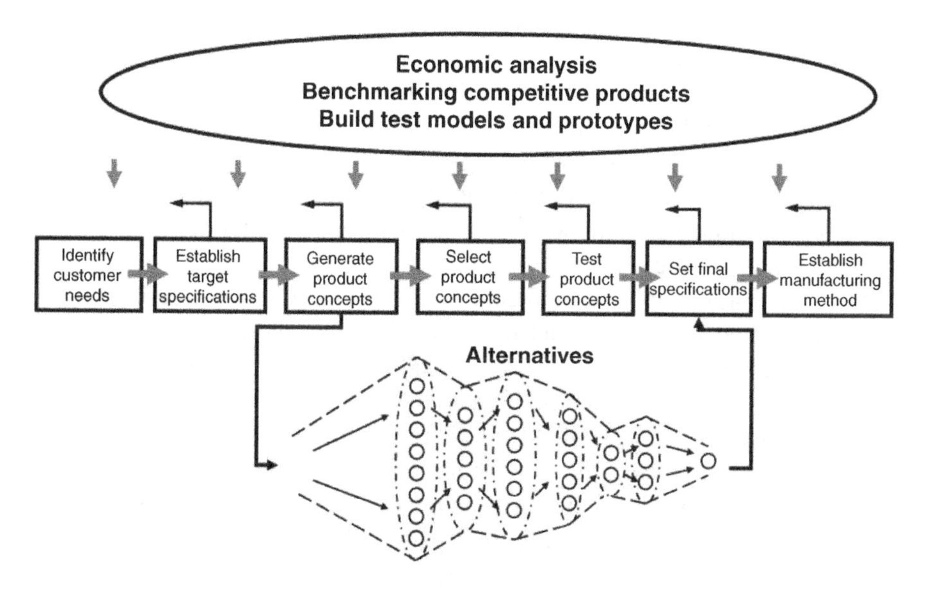

Figure 2.12 *The iterative development process according to Ulrich and Eppinger [17].*

At this stage, it is difficult to assess the realism of the specifications, but target values are set for the attributes qualitatively or in ranges. The target specification directs the activity in the third step, the concept generation, where a large number of concepts are generated based on known techniques and material properties. Since it is unrealistic to test and prototype all concepts, these are screened toward criteria derived from the target specification and are subsequently scored, or assessed, according to metrics (Figure 2.12). With one or two concepts remaining after the scoring assessment, prototypes are built and tested in thorough testing programs [18].

Of particular interest is to integrate the manufacturing requirements into the design, the so-called design for manufacturing (DfM) [19] (see also Chapter 7). The aim is to ensure that those design concepts that are more feasible for manufacturing become favored in the screening and assessment process. For biotechnological and biomechatronic products, it is important that the biological attributes and criteria of relevance for manufacturing are carefully considered when they are listed in the target specification. These criteria should also be ranked high when scoring the manufacturing alternatives. Thus, the biological experts of the team must identify the biology-related manufacturing requirements early, and, if possible, also anticipate their cost effects.

2.8 CONCLUDING REMARKS

The theory description in this chapter has brought up the essential principles of modern design methodology. As emphasized throughout the chapter, the theory and its methodology have predominantly been used for mechanical product design and development. In this book, we will further develop and apply the theory on products with a significant biotechnological content. We do this since we are convinced that biotechnology products can be developed and designed more easily and with more success for their intended purposes by using the methodology.

In Chapter 4, the applied biomechatronic methodology is described from the mechatronic perspective outlined above. Thus, reasoning and analysis based on transformation of energy, materials, and signals within the systems and functions are applied on the biotechnology products.

The Hubka terminology and symbolism are amply taken advantage of for describing and analyzing the biotechnology design problems. But we also bring into this procedures that are based on the systematic thinking and reasoning as presented in several other parts of this theory.

In order to make the methodology more adaptive to the added complexity of biology and biotechnology, we support the utilization of the methodology with additional design tools (see Chapter 4). These tools are then used in the cases presented in the preceding chapters.

Before entering into these hands-on biomechatronic tools, a brief account of the prerequisites the biological systems impose on systematic design due to its inherent complexity (Chapter 3) is given.

REFERENCES

1. Pahl, G., Beitz, W. (1996) *Engineering Design: A Systematic Approach.* Springer, Berlin.
2. Pahl, G., Beitz, W., Feldhusen, J., Grote, K.H. (2007) *Engineering Design*, 3rd edition. Springer, New York.
3. Kesselring, F. (1954) *Technische Kompositionslehre.* Springer, Berlin.
4. Tschochner, H. (1954) *Konstruieren und Gestalten.* Girardet, Essen.
5. Niemann, G. (1975) *Maschinenelemente Bd. 1.* Springer, Berlin.
6. Leyer, A. (1971) *Maschinenkonstruktionslehre 1–6.* Birkhäuser, Reihe, Basel.
7. Hansen, F. (1965) *Konstruktionssystematik.* VEB Verlag, Berlin.
8. Koller, R. (1994) Konstruktionslehre für der Maschinenbau. Grundlagen, Arbeitsschritte, Prinziplösungen. Springer, Berlin.
9. Hubka, V., Andreasen, M.M., Eder, W.E. (1988) *Practical Studies in Systematic Design.* Butterworths & Co. Ltd, London, UK.
10. Hubka, V., Eder, W.E. (1988) *Theory of Technical Systems: A Total Concept Theory for Engineering Design.* Springer, Berlin.
11. Hubka, V., Eder, W.E. (1996) *Design Science.* Springer, Berlin.
12. Ehrenspiel, K. (1985) *Kostengünstig Konstruieren.* Springer, Berlin.
13. Andresen, M.M., Hein, L. (1987) *Integrated Product Development.* Springer, Berlin.
14. Archer, L.B. (1985) The implications for the study of design methods of recent developments in neighbouring disciplines. In: *Proceedings of ICCD,* Heurista, Zurich.
15. Roozenburg, N.F.M., Eekels, J. (1996) *Product Design: Fundamentals and Methods.* Wiley, Chichester.
16. Ullman, D.G. (2003) *The Mechanical Design Process*, 3rd edition. McGraw-Hill, New York.
17. Ulrich, K.T., Eppinger, S.D. (2008) *Product Design and Development*, 4th edition. McGraw-Hill, New York.
18. Wall, M.B., Ulrich, K.T., Flowers, W.C. (1992) Evaluating prototyping technologies for product design. *Res. Eng. Des.* 3,163–177.
19. Poli, C. (2001) *Engineering Design and Design for Manufacturing: A Structural Approach.* Butterworth-Heinemann.

3

Biotechnology and Mechatronic Design

This chapter emphasizes the characteristics of the biological components of the biomechatronic systems and products. In order to systematically integrate the biological parts of a technical design we must have an accurate description of their characters. In the biological world these characters can vary considerably for each part depending on conditions. This is much owing not only to biochemical composition but also to architecture and microstructure of cells. Biomolecular structures or larger biological structures such as organs or individuals or whole biotopes have a great impact on the systematic design. These complexities are added to the extensive mechatronic prerequisites outline in Chapter 2 and by that impose new considerations and complex facts to the existing complexities of the technical system. This chapter systemizes the biological parts and highlights the biological features that are critical to consider.

3.1 TRANSDUCTION OF THE BIOLOGICAL SCIENCE INTO BIOTECHNOLOGY

Bringing the biological science into daily life applications has been done since ancient days. As mentioned in the introduction already, the Mesopotamians

Biomechatronic Design in Biotechnology: A Methodology for Development of Biotechnological Products, First Edition. Carl-Fredrik Mandenius and Mats Björkman.

fermented wines and the Romans produced olive oil in industrial refineries. In applied medicine, technical devices were used by the ancient Egyptian surgeons. In late eighteenth century, pharmaceutical technique for vaccination of small pox was begun and agricultural methods became a technology with harvesting machines. In the nineteenth century, industrial biotechnology set up protein enzyme production in fermenters with selected cultures. During the twentieth century, we were flooded with numerous biotechnology applications such as antibiotic manufacture, monoclonal antibodies for fast diagnosis, polio vaccines, recombinant insulin, and personal medicine with DNA mapping.

These examples all describe how biology interacts with mechanics and where the mechanics is the enabling means that allows the force of the biological systems to be exploited.

In a wider sense, mechanical systems can be expanded to hydrodynamics, fluid mechanics, reaction chemistry, thermodynamics, electronics, and other sciences related to engineering, which in one way or another can interact with biology in order to shape a technical system. In the introduction we coined the term "biomechatronics" to demark developments that are based on these preconditions.

In spite of this development, the design science and design theory we described in the previous chapter has had limited impact on the development of biology and biologically derived machines and system products. We quoted Gerald Pahl and Wolfgang Beitz who had stated:

> The main task of engineers is to apply their scientific and engineering knowledge to the solution of technical problems, and then to optimize those solutions within the requirements and constraints set by materials, technological, economic, legal, environmental and human-related considerations. [1]

Obviously, these engineers may be bioengineers, the materials may be microbial or human cells and biomolecules, and the legal aspects are covered by regulatory and ethical restrictions.

The mechatronic design theory is very compatible with biology although it has seldom been utilized in bioengineering training and, as a consequence, probably not very much in industrial practice.

We have shown in a few recent papers this possibility and how it can be applied to several of the mechatronic methods [1–5]. In this chapter, we further discuss the biological part itself and how it should be better merged with the mechatronic design methodology.

Here, we will try to further understand how biological principles are turned into technical products where the biology drives the invention/design concept.

3.2 BIOLOGICAL SCIENCES AND THEIR APPLICATIONS

In Chapter 2, we stated that all technical systems are based on the principle of energy conversion, transfer of materials, and signal transmission. Biological systems are not exceptions from this rule. On the contrary, it is difficult to find any biological system that does not obey this principle.

It is in this context we remind the reader of the range of the fields of biology and its diverse applications. Table 3.1 lists main areas of the biological

TABLE 3.1 Areas of Biology and Biology-Related Sciences with High Impact on Daily Life

Main Areas	Subareas	Impact on Daily Life
Zoology	Entomology	Control of noxious insects
	Vertebrate biology	Veterinary care
	Developmental biology	Stem cell manufacture for regenerative medicine
Botany	Forestry sciences	Pulp and paper industry
	Agricultural sciences	Farming
	Horticultural science	Landscape gardening
Microbiology	Microbial physiology	Biotechnology processes
	Mycology	Food applications of fungi
	Bacteriology	Infectious diseases and antibiotics
	Virology	Influenza and vaccines
Cell biology	Membrane biology	Drug uptake in intestine
	Embryology	Fetal diagnostics, stem cells
	Nuclear biology	Nuclear medicine
	Physiology	Athletic medicine
Toxicology	Developmental toxicology	Control of toxicants
	Organ-specific toxicology	Safety of drugs
	Environmental toxicology	Safe environment
Ethology		Farming, piggeries
Environmental science		Bioremediation of land
Molecular biology	Nucleic acid biology	Gene technology, forensic DNA analysis
	Glycobiology	Vaccines
	Protein biology	Diabetic treatment
Systems biology	Genomics	Personalized medicine
	Proteomics	Production of food proteins
	Metabolomics	Production of vitamins
Ecology		Agricultural farming
Biomedicine	Clinical medicine	Hospital testing
Biochemistry	Protein biochemistry	Detergent enzymes
	Molecular biochemistry	Antibiotics production
	Immunology	Diagnostics
Biological analysis		Biosensors
		Microscopic techniques
		Microarrays
		Clinical analysis

sciences with their subareas. As easily could be noted, this list is far from complete. Its purpose is just to exemplify how basic biological sciences influence our lives through applications we daily need.

Actually, it is difficult to find any branch of the biological science that does not result in practical applications. Man's ingenuity to exploit the inborn capacity of biological systems to convert things from the energy, material, or signals is striking. An incredible amount of possible combinations to turn these biological capacities into machines or processes to the benefit of society have been exploited since a long time ago.

Some areas of biology, such as evolutionary biology, taxonomy in zoology and botany, are still largely serving to further our understanding of the cosmos around us, from a historical perspective to the present. The impact on philosophy and even religion is undeniable and by that, of course, is an applied science in some sense – although it does not result in products or services.

The list of Table 3.1 shows another interesting feature. A substantial part of the applications are associated with the microscale or nanoscale biology – thus, what goes on inside a cell, a virus, or a DNA strand has today generated numerous practical applications in a vast number of areas such as medical diagnostics, forensic testing, and environmental analysis.

Probably most people identify applications of biology with medical care and drugs or with agriculture and farming due to visibility of these services and products. Other examples are more disguised. Few notice enzyme additives in detergents and contact lens liquids. The use of enzymes for the manufacture of marmalades and jams most people are not even aware of, although we would immediately complain on the product if the right taste and texture were missing.

Besides medicine, biology sciences have their greatest role in agriculture and food technology with the microbial products. These products link biology directly to manufacturing and industrial activities.

This makes the engineering aspects of biology apparent as will be discussed in more details below.

The biology areas are to a large extent overlapping and not discretely separated as the table may imply. In Figure 3.1, the interrelationships between the main areas and the applications are perhaps more apparent. This understanding is important. Biology can seldom be understood from a single subarea, but must take into account the whole biological system, as is so prominent in systems biology with its applications in genomics, proteomics, metabolomics, and transcriptomics. The need for seeing the whole picture, and understanding how the subparts are connected, has resulted in an additional understanding that often from the design engineering

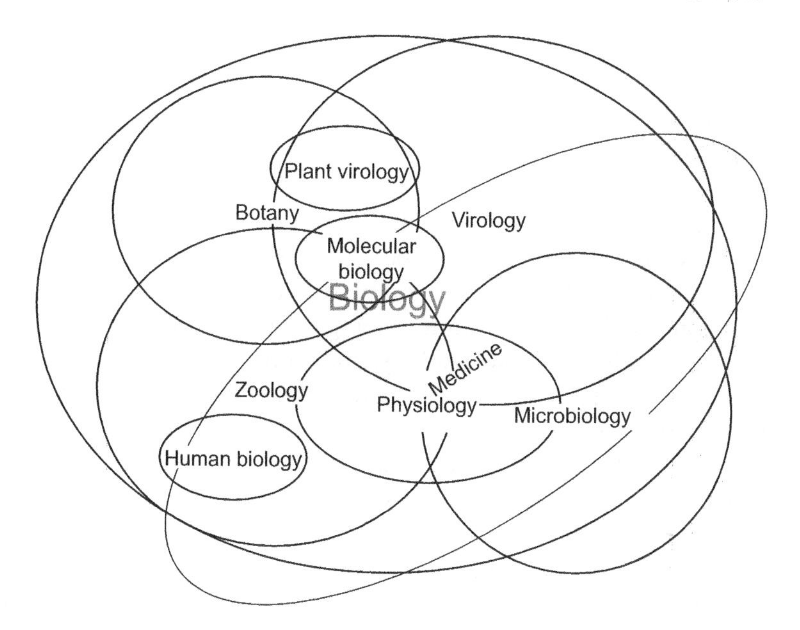

Figure 3.1 *Overlapping biological competences.*

perspective is critical for the function and operation of technical systems. Without knowing how the genome interacts with the metabolome and transcriptome, our ability to exploit the cellular machinery is severely impeded. The functionality of systems is a critical characteristic for the operation of the microcosm of the cell. It is also very critical for the functionality of the technical systems on a larger scale in any manmade machine, including those sprung from biotechnology inventions.

Furthermore, the knowledge about biological microscale world is needed also for analysis in the design of technical systems where interactions are pivotal for interpreting effects and interactions. Thus, genuine biological interactions that we predominantly have studied inside or between parts of biological systems need to be understood from an external view – from how they interact with nonbiological entities. These nonbiological parts could, as mentioned, be mechanical, electronic, optical, thermal, or chemical interactions. But it also includes social interactions, that is, how people interact with the systems – influence them or are influenced by them. Without this understanding, a complete integration of systems cannot be realized. When biological system is an integral part of this whole technical system, it is necessary to understand the underlying interactions and the possible effects on the whole technical system if this is going to be operable and meet the design intentions.

3.3 BIOTECHNOLOGY AND BIOENGINEERING

Thus, the biology applications utilize the capacity of the biological systems. Since the past 40–50 years we have been naming this biotechnology or bioengineering, depending on if we want to emphasize the technology aspect or the engineering aspect of the application. Today, we encompass in the concept of biotechnology a wide scope of applications: molecular biotechnology, nanobiotechnology, agricultural biotechnology, marine biotechnology, gene biotechnology, food biotechnology, forest biotechnology, space biotechnology, analytical biotechnology, and so on. In its widest definition, biotechnology is all application areas where biology is applied for technical or methodological purposes.

The concept of bioengineering has developed into an engineering science approach to the biology. This includes for example biochemical engineering, biomedical engineering, food engineering, genetic engineering, plant culture engineering, bioreaction engineering, biothermal engineering, bioanalytical engineering, biophysical engineering, biomaterials engineering, and so on. The bioengineering aspect is here related to the technique of carrying out the technology, how to design an engineering system, for example, a biochemical reaction, a biomedical device, a genetic construct, or a bioanalytical instrument. The bioengineering science relies much on general engineering mathematics where energy and mass balances are used and kinetics and dynamics are described for the systems. Typically, this is related to conversions of materials, energy, and signals.

Key features of bioengineering are production of materials: cells, such as fodder products, stem cells, or cell lines; proteins such as insulin, monoclonal antibodies, vaccines, or restriction enzymes; and metabolites, such as antibiotics, bioethanol, or biodiesel. In biotechnology, we often exploit the genetic capacity of the biological system, either as it is created by Nature or by using genetic engineering methods to improve it.

Other key features are observed when, for example, we use a biological system for producing information or signals through its capacity to recognize. What normally then is recognized are analytes, such as glucose or HIV in blood, environmental toxins, or complementary DNA or RNA sequences in forensic analysis.

Yet another key feature is seen when we try to rebuild with technical means the functions of a human organ. This could be exemplified by artificial liver devices and artificial heart veins.

All the application examples in this book (Chapters 5–13) bring the mechatronic design theory into these types of biotechnology design tasks.

TABLE 3.2 Examples of Biotechnology Applications, Their Biological Basis, and Their Mechatronic Interfaces

Biotechnology Application	Biological Basis	Mechatronic Interface
Production process for antibiotic drugs	Microbial metabolism and growth of fungi	Bioprocess manufacturing equipment
Antibody production	Hybridoma cell culture	Highly contained manufacturing plant
Biosensor device for immunoanalysis	Monoclonal antibody	Detection system and fluidic device
Protein purification system	Affinity ligands immobilized on a chromatographic media	Column with adapters, flow distributors, pumps, and effluent detectors
Blood glucose analyzer	Glucose-specific enzymes	Electrodes or optical transducers
cDNA microarrays	Nucleotide sequences, tagging molecules	Array chips, detection systems, software identification
MALDI-TOF mass spectrometry	Database for protein fragment weights	Sample preparation matrix. Qualified mass spectrometry techniques. Software data evaluation
Cell culture bioreactor	Compatibility with mammalian cells and culture media	Sterilizable stirred bioreactor with auxiliary equipment
Production of stem cell-derived organ cells	Human embryonic stem cells and differentiation factors	Robotics, analytic instruments, and controlled environment in incubators
Artificial liver	Functional hepatic culture and culture media	Multicompartment device for cells and media
Ulcer diagnostics	*Helicobacter* conversion in stomach of nuclide substrate	Geiger–Mueller detector device. Tablet with nuclide. Breathe card

Table 3.2 presents a list of the above-mentioned or similar biotechnology applications and denotes the type of capacity that the biological system provides and what interfaces are used to the mechatronic systems.

The three above-mentioned categories of key features can be distinguished in the table. One is the production of biological cells or molecules. The cellular capacity to carry out biosynthesis results in antibiotics, hybridoma cells with monoclonal antibodies, stem cells, and stem cell-derived organ cells. In these cases, the mechatronic interface achieves the containment and environment for this to happen. This is of course a complex technical design task. To try to do this without systematic consideration of interaction between biological systems and other technical systems is impossible.

The category of measuring signals is exemplified in the table with respect to blood glucose, occurrence of *Helicobacter* infection in stomach, DNA samples, protein fragments in MALDI-TOF mass spectrometry, and immunobiosensing. The biological systems provide the function of recognition of the biomolecules or cells of interest. The key feature of the mechatronic interface is the detection and interpretation of the signals.

The third category is exemplified only with the artificial liver device. This can be expanded with other organs such as kidney, pancreas, veins, blood cells in the circulation, and so on. The biosystem is there to imitate the natural organ for various applications. The functions need to imitate the human organ accurately or at least as accurate as possible if it concerns with surgery, regenerative medicine, or toxicity test of drug on the artificial liver. To achieve such as biological system is a challenge. The mechatronic interfaces are as a consequence also demanding.

3.4 APPLYING MECHATRONIC THEORY TO BIOTECHNOLOGY: BIOMECHATRONICS

In the present design methodology, as presented in Chapter 2, all technical systems are regarded equal. Thus, either it is a technical system derived from mechanics, electronics, optics, or chemistry it belongs to or it is a subsystem of the general technical system.

In the approach taken in this book, the biological systems are always analyzed separately from other technical systems. This is motivated by the fact that the biological systems are enormously complex and interconnected. Their impacts onto the other systems of the design are therefore significant.

Moreover, the biological systems are also complicated by the fact that they themselves are sometimes consumed or reproduced. For example, in a bioprocess we inoculate the process with a biological system, for example, a microbial cell. Thus, it is the material that is added to the process and converted. But the biological functions of this input biological material are able to reproduce the input into new materials with the same functions. This is a manifestation of the dogma of biology – sometimes referred to as the first principle of biology [1]:

$$\text{Cells} + \text{Raw materials (nutrients)} \rightarrow \text{New cells} + \text{Side products} + \text{Remains} + \text{Energy}$$

This condition makes the biological systems different from most other technical systems. Chemical systems are consumed and do not reproduce; mechanical and electric systems behave in the same way. Of course, biological systems need materials and energy and signals to reproduce – but the functions, or program code, are embedded into the biological materials and exert their effects on these materials. This is a unique feature that justifies the special treatment of the biological systems. The purpose to do so is that we shall study the interactions of the biological systems thoroughly and systematically versus the other technical systems, as well as other systems of the design.

Furthermore, systems biology, often regarded as a new scientific branch of biology, has substantial influences on medicine, biotechnology, and environmental biology. The knowledge about systems biology is now growing significantly owing to access to microanalytical tools and extended software power. In Chapter 8, we also apply these tools on the design concept.

Throughout this book we use the term biological systems, symbolically written as ΣBioS [5–7]. Consequently, it is one of the key features for describing the biomechatronic approach. This term shall encompass the key functions of the biological system in the systematic design efforts.

As a consequence of the design science framework outlined in Chapter 2, it seems adequate to assign ΣBioS a strict definition that allows energy, materials, and signals to be analyzed. Moreover, the subsystems of the ΣBioS should be defined from a functional perspective where the biological effects and interactions become apparent. Figure 3.2 illustrates this from a cellular perspective using a symbolic representation adopted from metabolic maps and biochemical reaction schemes. These are the intracellular metabolic pathways, the converting enzymes, the coenzymes necessary to support the enzymatic steps, the transport proteins and active transport/cotransport shunts/channels, the transcription and translation machinery, the cell division cycle, and the cellular growth. These functions and biomolecules can be

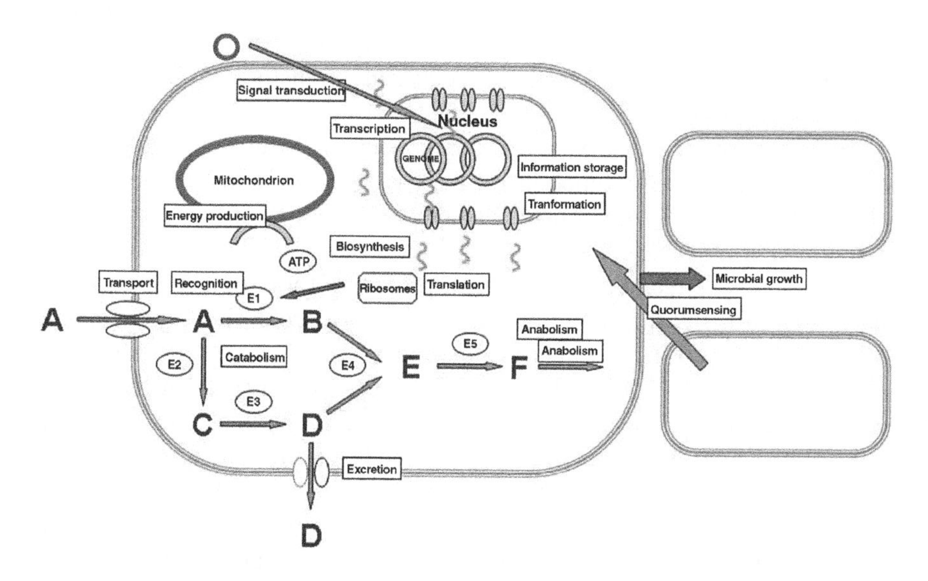

Figure 3.2 Subfunctions of a biological system (white labels).

ranged into a hierarchical relationship of subsystems and subfunctions. Together they form ΣBioS. The rationale for doing this is to open up and desiccate the interactions on a more detailed level. The interactions can be not only within the ΣBioS but also between the other systems and subsystems.

The functions human systems (ΣHuS) are in most cases clearly distinguished from ΣBioS by social or individual conditions. Despite being by necessity a biological system, ΣHuS are characterized by actions carried out by humans in their professional situation as operators, teachers, service men, doctors, and engineers. Conversely, the functions of a product may affect the function and capacities of ΣHuS, for example, the ergonomy of a product's design exert effects on the operator of the product or the physical form of a dentist's drill affects his ability to operate it with precision. The design solution shall consequently try to minimize such effects. In some instances, the distinction between ΣBioS and ΣHuS may be difficult to define. In the example in Chapter 7, a human/patient carries a stomach ulcer bacterial culture. The patient's body (intestinal tract/stomach) interacts with the culture and the culture interacts with a reagent to form a conversion product the technical system can record. In this example, the HuS is the patient, the culture the ΣBioS and the subsystems of the bacterium, that is, the transformations taking place inside the bacterium, that is, as shown in Figure 3.2.

The simple illustration of the functions in the cell in Figure 3.2 shows the essence of the biological processes. *Materials* (A, B, C, etc.) are transformed into the biological system with the help of *energy* (ATP that drives the biochemical transformations) and enzymes that execute the transformations of the materials from *signals* collected from the code of genome (DNA and mRNA) and the environment (signaling molecules and quorum sensing effectors). This is translated further by materials' transformation (tRNA and ribosomes) into functional enzymes.

The effects of the other parts and components of a (bio)-technical design will or could change the functions of the biological systems on the cellular or biomolecular levels. Of course, also the subfunctions of the ΣBioS affect each other.

The systems of technical equipment, process information, management, and regulations, and the environmental factors that are difficult to anticipate and control all exert their effects on functional biology. Figure 3.3 illustrates this. In the upper left circle, technical functions are carried out by certain gears/devices. In the lower left circle, information is handled, here illustrated by integrated circuits, by a mobile telephone and measurement instrument that can record and deliver information.

The upper right circle shows laws and regulations that restrict the technical systems' sometimes larger potential. This also includes goal/target values to be communicated to the personnel (HuS).

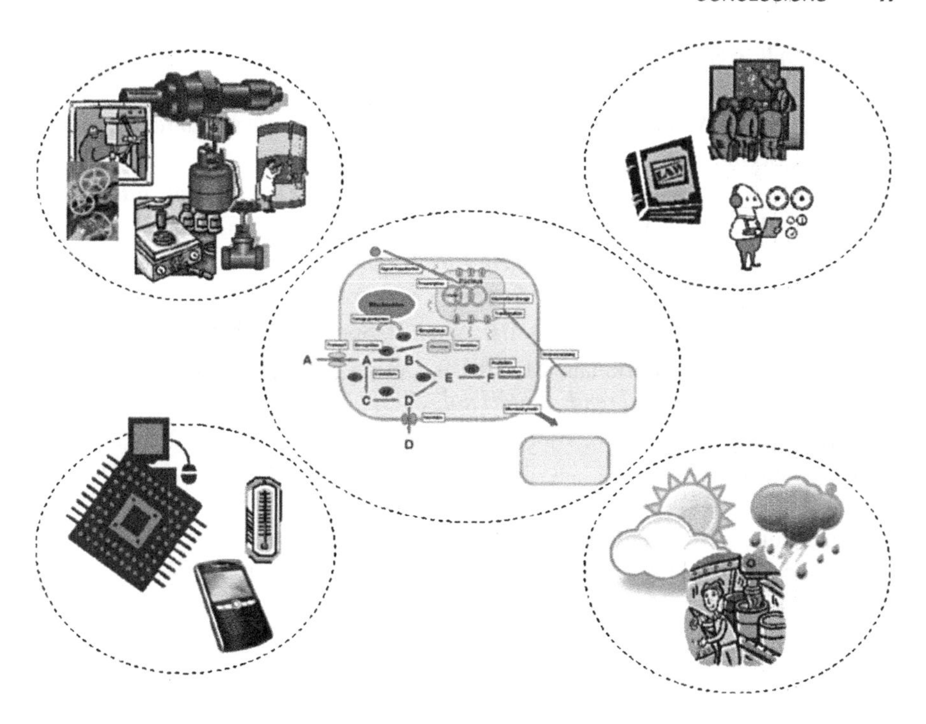

Figure 3.3 The functions of biological system are affected by the surrounding technical systems of the design task. But effects are also from information systems, management systems, laws, people involved interfering with the design, and the physical environment.

The lower right circle represents the active environment (AEnv). This comprises "disturbances" that interfere with the functions. AEnv is unpredictable. This is typical for a biological system, at least in the preunderstanding phase. Once new relationships are revealed, by true discovery or at least methods to predict behaviors, the AEnv can enter into the ΣBioS as a subsystem that exhibits mechanistic responses.

3.5 CONCLUSIONS

This chapter has described how materials, energy, and signals form the key elements of all biological systems and how these elements are intrinsically interconnected with the functions of the biological systems. A complicating aspect of biological systems is that they at the same time also transform themselves. How the other systems in a design affect and interact with the biological systems is the main task of the mechatronic design theory in biotechnology. The rest of this book is devoted to that subject.

REFERENCES

1. Pahl, G., Beitz, W. (1996) *Engineering Design, A Systematic Approach*, Springer-Verlag, Berlin.
2. Ulrich, K.T., Eppinger, S.D. (2008) *Product Design and Development*, 4th edition, McGraw-Hill, New York.
3. Hubka, V., Eder, W.E. (1988) *Theory of Technical Systems, A Total Concept Theory for Engineering Design*, Springer-Verlag, Berlin.
4. Hubka, V., Eder, W.E. (1996) *Design Science*, Springer-Verlag, Berlin.
5. Mandenius, C.F., Björkman, M. (2010) Design principles for biotechnology product development. *Trend. Biotechnol.* 28(5), 230–236.
6. Derelöv, M., Detterfelt, J., Björkman M., Mandenius C.F. (2008) Engineering design methodology for bio-mechatronic products. *Biotechnol. Prog.* 24, 232–244.
7. Mandenius, C.F., Derelöv, M., Detterfelt, J., Björkman, M. (2007) Process analytical technology and design science. *Eur. Pharm. Rev.* 12, 74–80.

4

Methodology for Utilization of Mechatronic Design Tools

This chapter describes a biomechatronic methodology for adapting and utilizing design methodologies and tools from the mechatronic design area to design biomechatronic products and systems. The methodology and tools are used in the biotechnology design applications described in the following chapters. Here, we explain how the design tools are best taken advantage of in general. It is explained when, how, and why they should be used. In particular, the biotechnology aspect is highlighted based on the views presented in Chapter 3.

4.1 IDEA OF APPLYING THE MECHATRONIC DESIGN TOOLS

The generic mechatronic design tools we describe and use as a basis in this chapter have all been previously presented by several authors [1–5]. Some of these authors also describe thoroughly how to use the tools in industrial design in a practical and optimal way in order to achieve a good design result.

Our approach in this book is to take established design tools from the mechatronic area and to modify, develop, and adapt them for application in

Biomechatronic Design in Biotechnology: A Methodology for Development of Biotechnological Products, First Edition. Carl-Fredrik Mandenius and Mats Björkman.

biomechatronic area. Furthermore, it is to bring them together in a partly new working scheme and methodology bringing the biotechnology aspects of the design close to the mechatronic issues. Below we present the tools in the order they normally shall be applied in the methodology for design of biomechatronic products and systems. This is further made clear when reading the applications of Chapters 5–13.

In Figure 4.1, the design tools are shown in a workflow diagram that depicts the biomechatronic methodology that is based on mechatronic methodologies.

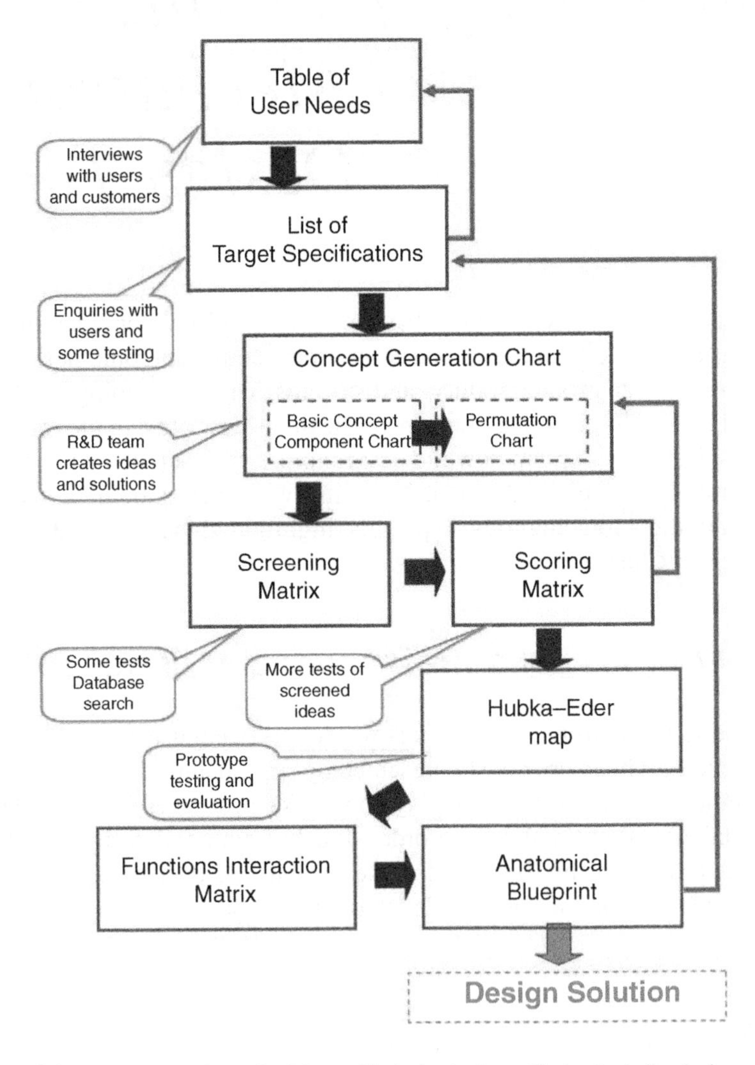

Figure 4.1 Biomechatronic methodology with design tools and their roles in the design workflow as used in the applications.

The structure of the workflow should not be taken as absolute. Special applications may motivate to deviate to some extent.

In most of the application examples, we begin the design work by defining the goal of the design. Once this is done, it is possible to list the *needs of the users*. These needs can then be transformed into concrete *specifications* for the needs, quite often with quantitative measures.

The specifications' different values give the boundaries for a design and based on these boundaries a number of *design concepts* are generated. It is often done by permuting combinations of functions.

The generated concepts are first systematically *screened* to remove those ideas that are not in line with the needs and specifications. The remaining concepts are thoroughly assessed by *scoring* by tests and other measurements.

The design alternatives with best scores are further studied. At this stage, a so-called *Hubka–Eder map* of the functions of the designed object is drawn. The map is a conceptual model-like approach where transformations and interactions in the design structure are lifted up to the design team's attention.

The interactions are then systematically evaluated in a *Functions Interaction Matrix* (*FIM*) that should show the most critical interactions.

Finally, this analysis of the critical interactions paves the way for setting an *Anatomical Blueprint* of the design. The anatomical blueprint goes into actual biological and mechatronic components and presents distinct and well-founded suggestions for realizing the conceptual design.

As Figure 4.1 tries to illuminate by reciprocal arrows, the design work is not a linear process in practice. It is a highly iterative process and without these iterations it is very difficult to achieve a proper design solution. It is also possible to omit design tools during a specific design task. This can be found in Chapters 5–13. All design tools are not utilized in all chapters. However, all the chapters follow the generic methodology order shown in Figure 4.1. With a sound utilization of design tools, recirculation/iterations, and evaluation of design concepts, the possibility to accomplish a feasible and optimal design solution is increased. Several of the steps in this methodology are similar to the Ulrich and Eppinger approach [1] that we have previously described [5].

4.2 TABLE OF USER NEEDS

Once the design mission is settled, the actual needs of the users must be confirmed. In many instances, the users are more or less obvious. Quite often, it is the same as the customers – those who are going to buy the designed product.

TABLE 4.1 Table of User Needs for the Design

Need	User Situation 1	User Situation 2	User Situation 3
Need 1	Attribute 1	Attribute 9	Attribute 17
Subneed 1	Attribute 2	Attribute 10	Attribute 18
Subneed 2	Attribute 3	Attribute 11	Attribute 19
Need 2	Attribute 4	Attribute 12	Attribute 20
Subneed 1	Attribute 5	Attribute 13	Attribute 21
Subneed 2	Attribute 6	Attribute 14	Attribute 22
Need 3	Attribute 7	Attribute 15	Attribute 23
And so on	Attribute 8	Attribute 16	Attribute 24

In other case, it can be more complex. In Chapter 7, we describe a device for stomach ulcer diagnosis. Here, users can be of at least three categories: patients, doctors, and nurses. However, the customer can often also be the local healthcare authority, which then probably is the paying customer.

The needs shall be compiled in a very thorough and structured way, neither excluding nor exaggerating needs. The method of collecting the information is preferably by interviews and enquiries, oral or written. The research material is analyzed thoroughly by the design team. Verification is important.

The *Table of User Needs* (see, for example, Table 4.1) could be structured into a few columns for specific situation, for example, groups of users or subapplications of the design mission. The rows in the table could be structured into groups of needs.

All application examples in this book (Chapters 5–13) have in the beginning of the chapter an analysis of the user needs. Normally, this is included in the *List of Target Specifications*.

4.3 LIST OF TARGET SPECIFICATIONS

The user needs can be transformed into concrete *specifications* for the needs, quite often but not always with quantitative measures (see, for example, Table 4.2). It is important not to exclude a need just because it is qualitative and cannot be quantified. The unit can be "Yes" or "No". The List of Target Specifications is the result of this analysis and transformation.

4.4 CONCEPT GENERATION CHART

The needs and target specifications' definitions and demarcations from the Table of User Needs and the List of Target Specifications provide a basis for generation of concepts.

TABLE 4.2 List of Target Specifications for the Design Mission (Example)

User Need → Metrics	Target Values	Units
Need 1 (user convenience)		
Subneed 1.1 (flexible use)	Value 1	No. of different analytes
Subneed 1.2 (time to change unit)	Value 2	Minutes
Subneed 1.3 (switch on time)	Value 3	Seconds
Subneed 1.4 (hand-held)	Value 3	Yes/No
Need 2 (Performance)		
Subneed 2.1 (Precision of work)	Value 5	%
Subneed 2.2 (response time)	Value 6	Seconds
Subneed 2.3 (detection limit)	Value 7	Part per million
Subneed 2.3 (operation range)	Value 8	Molar
Need 3 (User costs)		
Subneed 3.1 (instrument unit cost)	Value 9	EUR per unit
Subneed 3.2 (disposables)	Value 10	EUR per disposable
Subneed 3.3 (maintenance)	Value 11	EUR per year
Subneed 3.4 (labor cost)	Value 12	Hours per sample

This should be done from a functional reasoning. The reasoning is done on two levels: first functional description as generic as possible is done where one is trying to reduce the essence of the functions as much as imaginable. This is here referred to as the *Basic Concept Component Chart*. Once this has been settled, the components are permuted and certain design components can be added. By that, several alternatives are generated which we call *Permutation Chart*. In this treatise, we have put these two together in a *Concept Generation Chart*.

4.4.1 Basic Concept Component Chart

A Basic Concept Component Chart has normally 3–10 functional symbols depicting the different basic functions in the form of *basic concept components* and a collection of arrows or dotted squares that relate the functions in a minimalistic way to each other. The purpose of this is to keep the design mission on an abstract level in order to minimize the risk that the designer teams look too early into certain physical design solutions. Especially in the beginning of the design process, it is vital to be open-minded to many alternative solutions as long as certain key functions are maintained.

The exercise to create a Basic Concept Component Chart can be relatively simple if the necessary background is present. The mindset of the workflow in Figure 4.1 is that the needs and target specification span out the functional parameters and that this thinking already there creates the notions and sentiments of what should be possible to generate. However, there is a great possibility that this thinking also creates ideas for the detailed design. The risk

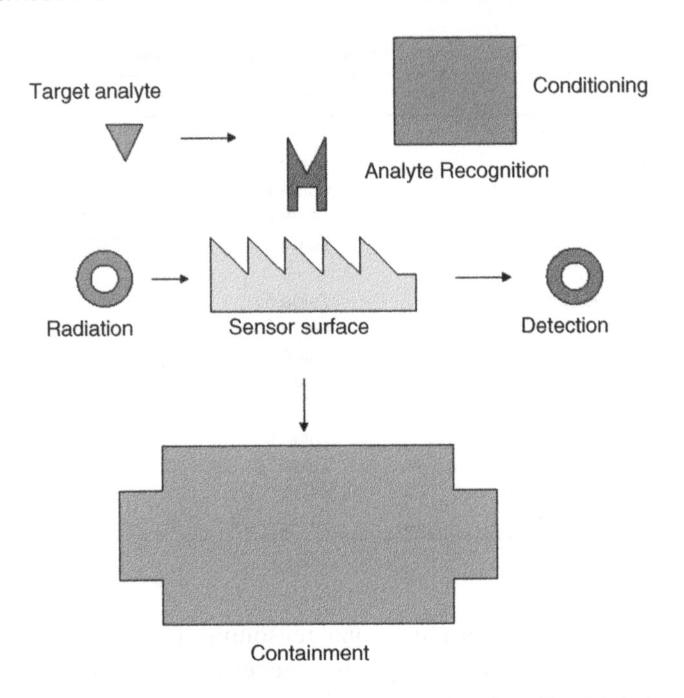

Figure 4.2 *Example of a Basic Concept Component Chart for a bioanalytical application.*

is that the design team focuses too early on these early ideas and thereby does not explore the full possibility of the existing design space. It is a vital role of the Basic Concept Component Chart to decrease that risk.

It is helpful if the symbols of the *Basic Concept Component Diagram* have some physical likeness of the function. In Figure 4.2, an example is shown from a bioanalytical application. Components of the diagram are typically functions for (1) detection and (2) radiation, (3) surface for analytical reaction, (4) analytes to be sensed, (5) analyte recognizing function, (6) conditioning media, and (7) some sort of containment of the system. The arrows indicate the expected course of actions or events.

4.4.2 Permutation Chart

In the following Permutation Chart, the Basic Concept Component Diagram is permuted. This means that the functions are recombined in as many possible configurations as the team can imagine, some looking promising from the beginning, others unlikely to work at all. Still the latter should not be excluded too early. To make the generation more creative, new functions or additives can be introduced and integrated into the permutations. In Figure 4.3, the permutation procedure is illustrated on the basis of the components in

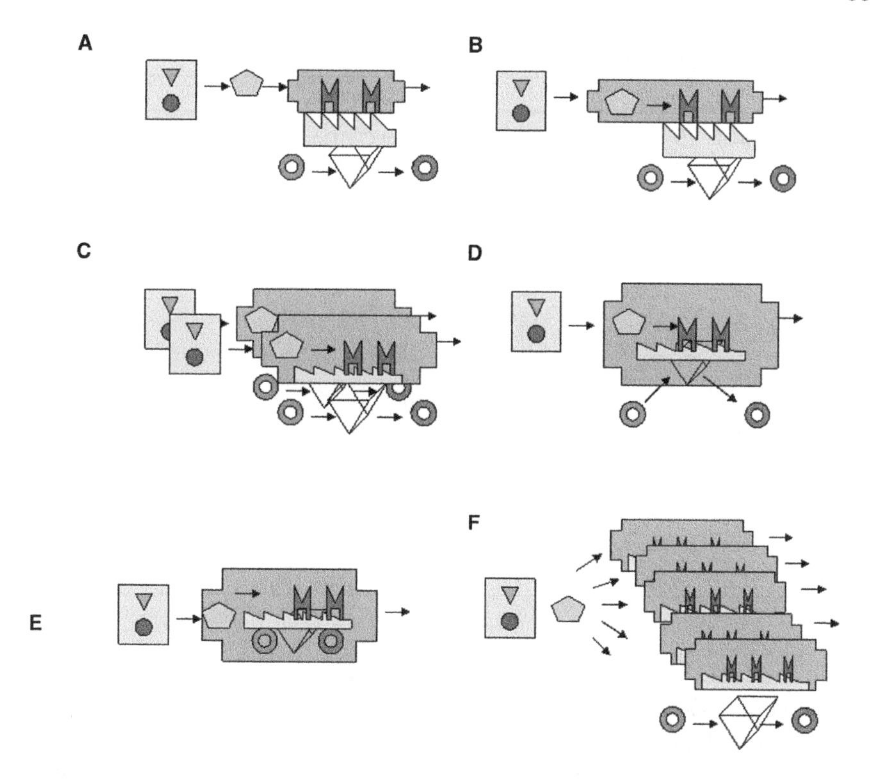

Figure 4.3 *Example of a Permutation Chart for the bioanalytical application presented in Figure 4.2. The basic components are permuted with a few new components added.*

Figure 4.2. The added functions are a beam splitter and an interfering analyte. The example is based on the surface plasmon resonance sensor described in Chapter 6 where this Permutation Chart is explained in more detail. It should be noted that many more than six concepts could be generated easily, especially when more functions are included and additional components added.

Other examples in this book where we apply the Concept Generation Charts are discussed in Chapters 8 (DNA microarray), 10 (chromatographic purification), 12 (stem cell manufacture), and 13 (a PAT system).

4.5 CONCEPT SCREENING MATRIX

The *Concept Screening Matrix* is directly adapted from the methodology suggested by Ulrich and Eppinger at MIT [1]. We apply it here as an immediate step after the concept permutation alternatives. These concepts are consequently screened versus the needs and specifications from the

TABLE 4.3 Concept Screening Matrix Example

	Concepts					
Selection Criteria	Concept A	Concept B	Concept C	Concept D	Concept E	Concept F
Need 1						
Subneed 1,1	0	+	−	+	+	+
Subneed 1.2	−	+	−	+	+	+
Need 2						
Subneed 2.1	−	+	0	+	−	+
Subneed 2.2	+	+	+	−	−	−
Need 3	−	+	0	+	+	+
Subneed 3.1	+	+	+	+	+	+
Subneed 3.2	−	+	0	+	+	+
Subneed 3.3	0	−	+	−	+	0
Sum +'s	2	7	3	6	6	6
Sum 0's	2	0	3	0	0	1
Sum −'s	−4	1	2	2	2	1
Net score	−2	6	1	4	4	5
Rank	**6**	**1**	**5**	**3**	**3**	**2**

Table and List. Screening means here a coarse assessment of the capacity of the concepts to achieve the targets. Only three levels are assessed (− , 0, +). The sum is calculated, which should result in ranking of the concepts. If it is difficult to distinguish the alternatives, further testing must complement the screening procedures. It can also be other form of information collection, for example, from databases or expert panels. Table 4.3 shows a typical screening matrix of the form we apply in Chapters 5–13.

4.6 CONCEPT SCORING MATRIX

The concepts highest ranked in the Concept Screening Matrix are further assessed in a *Concept Scoring Matrix* (see Table 4.4). Now fewer concepts are handled in the assessment making it feasible to carry out more bench and lab-scale testing and investigation of the capacities of the concepts. In the matrix can the target specification be refined or focused on what is considered to be decisive issues. The assessment values are now more precise and their importance allotted a weight. The assessment values are calculated. An Excel sheet is useful to facilitate the calculation. A new ranking is performed.

In this part, biological laboratory experiments can be envisaged as well as prototype building. It means that this will take more time to perform and be more costly. However, it will increase the precision and quality of the design work considerably.

TABLE 4.4 Concept Scoring Matrix Example

		Concept A		Concept C		Concept E	
Selection Criteria	Weight	Rating	Weighted Score	Rating	Weighted Score	Rating	Weighted Score
Flexible use	20						
Home use	2	1	2	5	10	8	16
Clinic	3	4	12	7	21	9	27
Performance	40						
Test time	10	5	50	6	60	8	80
Sensitivity	10	5	50	7	70	8	80
User convenience	25						
Easy training	5	1	5	1	5	9	45
Support	10	7	70	8	80	7	70
Test preparation	5	4	20	7	35	8	40
Healthcare integr.	5	6	30	6	30	8	40
Cost	30						
Investment	5	3	15	4	20	6	30
Consumable cost/test	10	8	80	6	60	7	70
Operation cost/test	15	5	75	5	75	7	105
Manufacturability	35						
Desktop unit	20	3	60	4	80	6	120
Consumables	25	2	50	6	150	5	125
Total score			438		695		848
Rank			**3**		**2**		**1**

4.7 HUBKA–EDER MAPPING

4.7.1 Overview Hubka–Eder Map

The Hubka–Eder theory for conceptual design was first presented in 1988 by Vladimir Hubka and Ernst Eder [3]. Figure 4.4 illustrates the principle. Its fundament is the transformation process (TrP) with three recurring phases: the preparation phase, the execution phase, and the finishing phase.

The TrP is fed by primary input operands (ΣOdIn) – those inputs of materials, energy, or anything else that will be directly transformed – and the secondary inputs (ΣSecIn) – those that are needed in the TrP, but not taken up by or in the final product. The primary output operands (ΣOdOut) are the expected results of the transformation (i.e., a pure protein, a produced recombinant protein, or an analytical result), and the secondary or residual outputs (ΣSecOut) are those that are not included in the main products (by-products or unused nutrients) [4].

To carry out the TrP, several technical systems (ΣTS) are required for realizing the phases of the TrP (Figure 4.4). In general terms, each TS causes a decisive effect on the TrP, and their functionality, or ability to cause the effect, is of vital interest for the design concept. Normally, the number of ΣTS is fairly

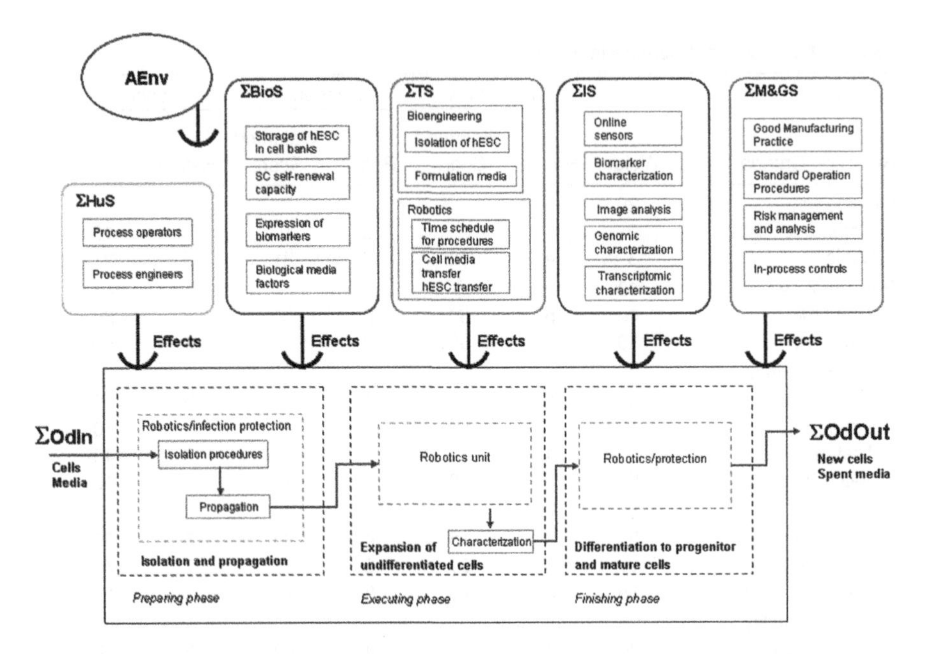

Figure 4.4 *Example of a Hubka–Eder map of the design functions. The example shows a manufacturing process for human embryonic stem cells where the embryonic cells (inputs) are differentiated and expanded to new cells (output). In the TrP, the processing phases are depicted. Above the TrP are the five systems that create the effects responsible for carrying out the transformation process. Included is also the AEnv that causes effects on the TrP and which are responsible for often unanticipated variations.*

large, but they can also be subdivided into smaller parts, subsystems, that simplify the analysis at the expense of enlarging the complexity of the ΣTS.

The TrP and the ΣTS normally require humans for operation. Usually, several individuals with specific expertise are involved. Together, they form the human systems (ΣHuS) and are interdependent, and exert their effects in unique ways. The ΣTS and ΣHuS are further supported by the information systems (ΣIS). These observe the TrP and the effects of the ΣTS. The ΣIS affect operation and decision making based on management and goal systems (ΣM&GS); that is, set procedures on how to proceed or behave in defined situations or according to established or regulatory approved standards.

Finally, the Hubka–Eder theory introduces the active environment (AEnv), which refers to the unpredictable variations that affect the TrP and the systems in an unexpected way. This is recognized in biotechnical systems as the "biological variation," but might occur in any technical systems (i.e., weather conditions).

It is possible and appropriate to include biological and biochemical systems (ΣBioS) to the Hubka–Eder theory [6, 7]. This could be justified especially for cells in biological media in which the interactions and effects on the TrP and the other systems described above deserve special attention and are intrinsically complex to disentangle. Analyzing these effects in both directions, how the ΣBioS are affected by the choice of ΣTS alternatives, for example, could be an important success factor for a biotechnology product design.

The overview Hubka–Eder map is found in all of the application chapters (Chapters 5–13).

4.7.2 Zoom-in Hubka–Eder Mapping

Quite often it is necessary to analyze a certain area in the Hubka–Eder map. This is here called a *Zoom-in Hubka–Eder Mapping*. Most applications are too complex to be possible to describe in sufficient details on one map if the critical design functions should be covered. Figure 4.5 illuminates this zooming-in of a certain region of the map compared to the map of Figure 4.4. For biotechnology applications, it is obviously the biological systems of the design that normally are particularly interesting to detail more. We see in this example how the biological subsystems are broken up into typical

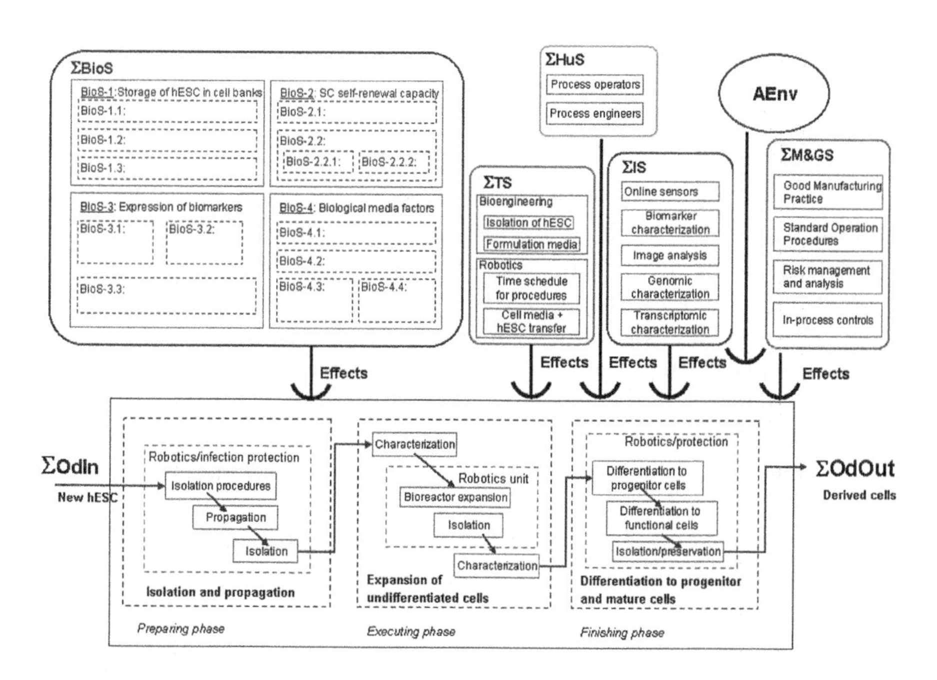

Figure 4.5 *Example of a zoom-in Hubka–Eder map of the design functions. Here, the functions and subfunctions of the ΣBioS systems are zoomed in. Also, the TrP functions are detailed more than in the example of Figure 4.4.*

biomolecular or cell biological functions. For example, the second biological system (BioS-2) stem cell self-renewal capacity is broken up into the subsystems[1] BioS-2.1 and BioS-2.2. The BioS-2.2 is further broken down into the subsystems[2] BioS-2.2.1 and BioS-2.2.2.

The zoom-in Hubka–Eder mapping is found in the applications of Chapters 6 (for detailed design of fluidics and immobilization), 9 (for detailed design of bioreactor parts), 10 (for special cases of micropurification and FVIII large-scale purification), and 13 (for details in the PAT systems design).

4.8 FUNCTIONS INTERACTION MATRIX

The Hubka–Eder mapping contains an essential description of the functionality of the design. But the most valuable contributions of the maps are in visualizing the interactions between the systems and how different subsystems can influence the functions of the transformation process. In this book, we use two types of interaction matrices: (1) those with the main objective of elucidating the interaction between the systems and the subsystems *per se* and (2) those that elucidate the interactions between the systems, on the one hand, and the transformation process steps on the other. These are discussed in brief in the following sections.

4.8.1 Functions Interaction Matrix for Systems and Subsystems

By placing the same system items, including subsystems, on both axes of the matrix (X and Y), each interaction can be identified at the different x/y-coordinates. Note that in a matrix the x–y pair appears twice, either on the right or on the left side of the diagonal axis of the matrix. These positions are not equal. The right side shows the x-item's effect on the y-item and left side shows the y-item's effect on the x-item. These effects may be very different in a design structure.

With FIM for systems and subsystems only the interactions between these are analyzed. Figure 4.6 shows an example from a bioreactor design case.

Figure 4.6 shows the matrix based on the ΣBioS and ΣTS of the Hubka–Eder map. The matrix has three levels of systems: the main systems and the first and second levels of subsystems. By not including all systems in the matrix, more space is left for more details by going further down in levels. This is something that is of great value for those systems that have the highest degree of complexity.

We apply this type of FIM in the design cases of blood glucose sensors (Chapter 5) and bioreactors (Chapter 9).

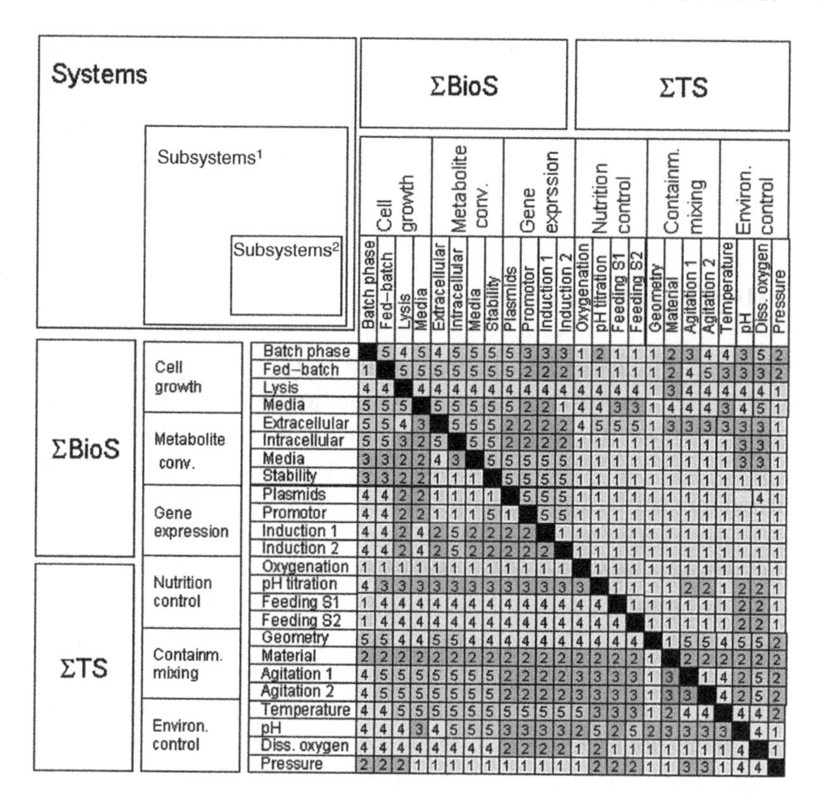

	Batch phase	Fed–batch	Lysis	Media	Extracellular	Intracellular	Media	Stability	Plasmids	Promotor	Induction 1	Induction 2	Oxygenation	pH titration	Feeding S1	Feeding S2	Geometry	Material	Agitation 1	Agitation 2	Temperature	pH	Diss. oxygen	Pressure
Batch phase		5	4	5	4	5	5	5	5	3	3	3	1	2	1	1	1	2	3	4	4	3	5	2
Fed–batch	1		5	5	5	5	5	5	5	2	2	2	1	1	1	1	1	2	4	5	3	3	3	2
Lysis	4	4		4	4	4	4	4	4	4	4	4	4	4	4	4	1	3	4	4	4	4	4	1
Media	5	5	5		5	5	5	5	5	2	2	1	4	4	3	3	1	4	4	4	3	4	5	1
Extracellular	5	5	4	3		5	5	5	2	2	2	2	4	5	5	5	1	3	3	3	3	3	3	1
Intracellular	5	5	3	2	5		5	5	2	2	2	2	1	1	1	1	1	1	1	1	1	3	3	1
Media	3	3	2	2	4	3		5	5	5	5	5	1	1	1	1	1	1	1	1	1	3	3	1
Stability	3	3	2	2	1	1	1		5	5	5	5	1	1	1	1	1	1	1	1	1	1	1	1
Plasmids	4	4	2	2	1	1	1	1		5	5	5	1	1	1	1	1	1	1	1	1		4	1
Promotor	4	4	2	2	1	1	1	5	1		5	5	1	1	1	1	1	1	1	1	1	1	1	1
Induction 1	4	4	2	4	2	5	2	2	2	2		1	1	1	1	1	1	1	1	1	1	1	1	1
Induction 2	4	4	2	4	2	5	2	2	2	2	2		1	1	1	1	1	1	1	1	1	1	1	1
Oxygenation	1	1	1	1	1	1	1	1	1	1	1	1		1	1	1	1	1	1	1	1	1	1	1
pH titration	4	3	3	3	3	3	3	3	3	3	3	3	3		1	1	1	1	2	2	1	2	2	1
Feeding S1	1	4	4	4	4	4	4	4	4	4	4	4	4	4		1	1	1	1	1	1	2	2	1
Feeding S2	1	4	4	4	4	4	4	4	4	4	4	4	4	4	4		1	1	1	1	1	2	2	1
Geometry	5	5	4	4	5	5	4	4	4	4	4	4	4	4	4	4		1	5	5	4	5	5	2
Material	2	2	2	2	2	2	2	2	2	2	2	2	2	2	2	2	1		2	2	2	2	2	2
Agitation 1	4	5	5	5	5	5	5	5	2	2	2	2	3	3	3	3	1	3		1	4	2	5	2
Agitation 2	4	5	5	5	5	5	5	5	2	2	2	2	3	3	3	3	1	3	3		4	2	5	2
Temperature	4	4	5	5	5	5	5	5	5	5	5	5	5	3	3	3	1	2	4	4		4	4	2
pH	4	4	4	3	4	5	5	5	3	3	3	3	2	5	2	5	2	3	3	3	3		4	1
Diss. oxygen	4	4	4	4	4	4	4	4	2	2	2	2	1	2	1	1	1	1	1	1	1	4		1
Pressure	2	2	2	1	1	1	1	1	1	1	1	1	1	2	2	2	1	1	3	3	1	4	4	

Figure 4.6 *Example of a Functions Interaction Matrix between biological and technical systems, subsystems¹ and subsystems². The following strengths of interactions are indicated: very strong (5), strong (4), intermediate (3), weak (2), very weak, or none (1).*

4.8.2 Functions Interaction Matrix for Systems and Transformation Process

It is also as valuable to investigate the interactions between the system functions and the functions in the three stages of the transformation process. On the main system function level, this is trivial, but on the subfunction levels it is not. In particular, a Functions Interaction Matrix for systems and transformation process becomes useful for comparing alternative choices of functions for a design.

We apply this type of FIM on blood surface plasmon resonance sensors (Chapter 6) and process analytical technology design (Chapter 13).

4.8.3 Design Structure Matrix

Sometimes in the literature, the concepts Design Structure Matrix and Design Interaction Matrix are used (see, for example, Refs [1, 2, 6]). In this book, we have modified these terms to Functions Interaction Matrix as described here to emphasize that the matrix is actually studying the interactions between conceptual functions and not interactions between the physical objects or components. The interactions between the physical objects or components should not be studied at this stage of the design work.

4.9 ANATOMICAL BLUEPRINT

The Functions Interaction Matrix for Systems and Transformation Process can highlight critical parts of the functional interactions with the transformation in the design. At this stage of the design process, it is time to focus on real physical objects and construction components, and this concerns both biological and mechatronic components. Thus, actual cells, proteins, antibodies, enzymes are to be considered. Mechatronic components in the technical systems such as pumps, actuators, detectors, valves, microprocessors, and so on should be brought forward and configured. For this, we use the term Anatomical Blueprint. This could be illustrated by a representation similar to the TrP scheme but where the components and component alternatives are included. In Figure 4.7, the functions are shown as tilted squares and the anatomical components as circular objects with a name related to the component. Figure 4.7 shows an example based on Figure 4.4 Hubka–Eder map example.

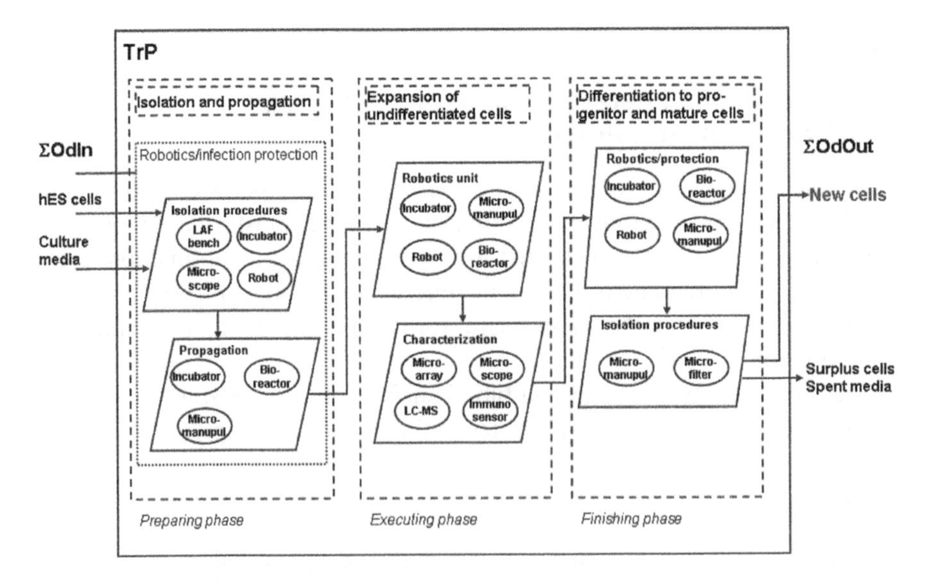

Figure 4.7 Example of an anatomical blueprint of the Hubka–Eder map in Figure 4.6.

4.10 CONCLUSIONS

This chapter has introduced the reader to a number of design tools for mechatronic design. The original tools are general, but they are adapted for biomechatronic applications in this chapter. The purpose of all the tools is to structure the design task. Furthermore, this chapter also describes an overall methodology for how to utilize these design tools in a structured manner in biomechatronic design applications. This will facilitate for the design team to organize the design work and to communicate.

All the design tools described are not necessary to be used for each design task. It is up to the designers to make the choice and to realize which tools are appropriate for a particular task.

In the design cases discussed in this book, the design methodology and tools are exploited and demonstrated.

REFERENCES

1. Ulrich, K.T., Eppinger, S.D. (2008) *Product Design and Development*, 4th edition, McGraw-Hill, New York.
2. Pahl, G., Beitz, W. (1996) *Engineering Design: A Systematic Approach*, Springer-Verlag, Berlin.
3. Hubka, V., Eder, W.E. (1988) *Theory of Technical Systems: A Total Concept Theory for Engineering Design*, Springer, Berlin.
4. Hubka, V., Eder, W.E. (1996) *Design Science*, Springer, Berlin.
5. Mandenius, C.F., Björkman, M. (2010) Mechatronics design principles for biotechnology product development. *Trend. Biotechnol.* 28(5), 230–236.
6. Hubka, V., Andreasen, M.M., Eder, W.E. (1988) *Practical Studies in Systematic Design*, Butterworths & Co. Ltd., London, UK.
7. Derelöv, M., Detterfelt, J., Björkman, M., Mandenius, C.F. (2008) Engineering design methodology for bio-mechatronic products. *Biotechnol. Prog.* 24, 232–244.
8. Mandenius, C.F., Derelöv, M., Detterfelt, J., Björkman, M. (2007) Process analytical technology and design science. *Eur. Pharm. Rev.* 12, 74–80.

Part II

Applications

5

Blood Glucose Sensors

This chapter discusses the design of blood glucose sensors. First, a basic background is given on clinical glucose testing and the diagnostic methods that are applied (Section 5.1). Then the needs of blood glucose analyzers are thoroughly specified (Section 5.2). Finally, these specified needs are used to develop new design solutions to blood glucose sensors (Sections 5.3 and 5.4).

5.1 BACKGROUND OF BLOOD GLUCOSE ANALYSIS

Blood glucose analysis is one of the most common tests in clinics, doctor's offices, and in homecare. The main utility for measuring glucose in blood is to support the treatment of diabetic patients, monitoring of patients under intensive care, and to detect states of hyper- or hypoglycemia.

Clinical glucose analysis is based on enzymatic conversion of the glucose in blood samples followed by detection using either spectrophotometric or electrode-based methods. Well-established laboratory methods are the Cobas-Bio centrifugal analyzer (Roche Diagnostics) and the YSI enzyme electrode method (YSI Model 23 AM, Yellow Spring Instruments, OH).

Biomechatronic Design in Biotechnology: A Methodology for Development of Biotechnological Products, First Edition. Carl-Fredrik Mandenius and Mats Björkman.

During the 1990s and 2000s, point-of-care testing was developed for small handheld glucose sensor devices based on different detection principles. In Figure 5.1, a few commercial glucose sensors are shown.

The HemoCue™ system (Hemocue AB, Sweden) uses a plastic cuvette that samples 5 μL of undiluted capillary blood from the finger. The cuvette

Test system	Principle	Reference
Hemo 	Photometric detection. Blood sample is treated by hemolysis with saponin, glucose dehydrogenase, diaphorase Whole blood samples Capillary	Hemocue AB (S)
Accu-Chek 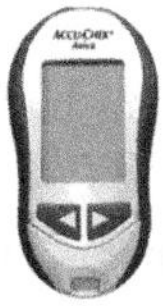	Amperometric detection. Blood glucose converted by glucose oxidase. Samples whole blood from capillary	Roche Diagnostics (D)
OneTouch 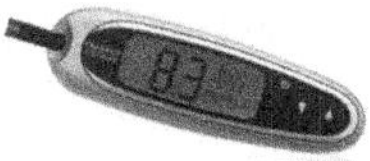	Photometric dection. Blood glucose converted by glucose oxidase and peroxidase. Samples capillary blood.	Lifespan (USA)
Precision Plus 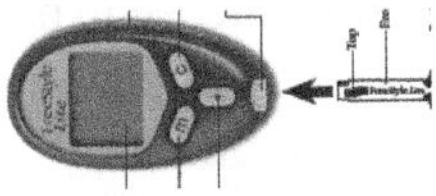	Amperometric detection Samples plasma from capillary	Medisense, Abbott Labs (D)
GlucoTouch 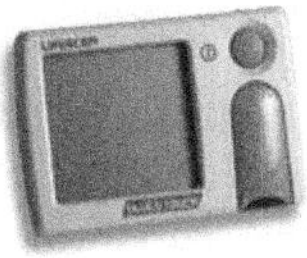	Photometric detection. Blood glucose converted by glucose oxidase and peroxidase. Samples capillary blood.	Johnson & Johnson (USA)

Figure 5.1 *Five commercial glucose sensors for point-of-care blood analysis. Reproduced with permissions from Hemocue, Roche Diagnostics, Lifespan, Abbott Labs, and Johnson & Johnson.*

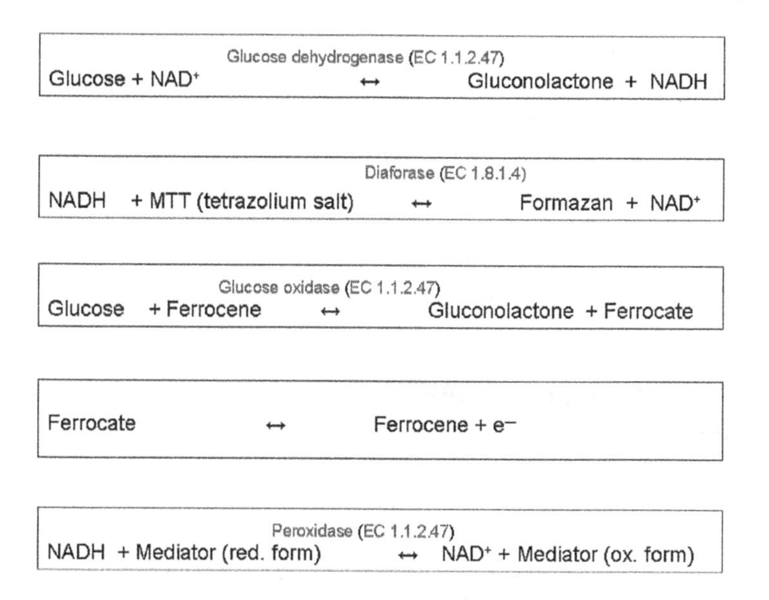

Figure 5.2 *Biochemical reactions in the glucose sensors shown in Figure 5.1.*

contains dry reagents and enzymes that disrupt the blood cells and convert the released blood glucose. The enzymatic conversion results in a colored compound that is detected in a handheld photometric detector unit. The enzyme reaction steps that take place in the sampling cuvette are shown in Figure 5.2.

The PrecisionPlus™ device from Abbott Ltd. (the United Kingdom) uses a 5 cm-long plastic stick that is brought into contact with the blood sample and then placed in a pocket-sized reader unit. The stick has a small spot with immobilized glucose oxidase that converts the blood glucose using ferrocene as a cofactor. When one glucose molecule is converted, one ferrocene is reduced to a ferrocate ion, the redox form of ferrocene (see Figure 5.2). The electrochemical reaction takes place directly on the enzyme spot of the stick where an electrical signal is generated and transmitted through wires embedded in the stick to the sensor reader unit.

In the Accu-Chek™ device from Roche Diagnostics (Germany), a variant of the amperometric principle is used but with other mediator than ferrocene.

The GlucoTouch™ device from Johnson & Johnson Inc. (the United States of America) is based on converting the blood glucose with glucose oxidase in a two-step reaction using peroxidase (Figure 5.2) whereby a colored reagent is formed and can be detected in the photometer unit.

All of the above-described sensor devices can analyze capillary, venous, or whole blood but with varying accuracy. A number of evaluations and comparisons of the correlation between the devices and the standard laboratory methods have been reported [1–11]. These reports all show certain variation in sensor performance. Typically, the variation is related to the category of patients tested and how and where sampling takes place, which, of course, significantly influence the accuracy and value of the comparison. Also, interferences by other components present in the patient's blood contribute to the variations observed. It seems an inevitable task that the design solutions should adequately address these variations in a creative way and try to reduce their effects.

5.2 SPECIFICATION OF NEEDS FOR BLOOD GLUCOSE ANALYSIS

Normally, a glucose analysis has a clear clinical purpose. In diabetic care, the purpose is to arrive at a correct decision on insulin dose. In intensive care, it is prevention of hypo- or hyperglycemic states. For a correct clinical decision, the reliability and precision of the measurement are critical parameters. It can be assumed that this depends not only on the measurement principle and the established design of the glucose sensor but also on the handling of the measurement method(s).

The World Health Organization (WHO) has issued recommendations on the use of point-of-care devices for diabetic treatment [12]. These recommendations put severe constraints on the use of glucose sensors and their design. Consequently, the WHO guidance should impact the design criteria and be visible in the specification of the design. In essence, the WHO bases its recommendation on the correlation with standard methods and reliability of results. Importantly, it also includes interferences and variations in populations.

Other requirements concern lot variation of sticks and reagents. Shelf-life should be acceptable for home use and doctor's offices throughput of tests.

Convenience for patients, the glucose sensors are probably the most used of homecare products, is obvious when considering the diabetic home treatment with insulin. Easiness of using and calibrating the device is a part of this. Time for test is integrated into the convenience concept and is also connected to the reliability of the test.

The software design of the sensor requires easy display of results and easy handling. Finally, several market-related issues could be taken care of when going through the needs for designing glucose sensor, such as the cost of device and consumables and service and support. The specification of the design is summarized in Table 5.1 listing the most prominent needs that should direct the design of a blood glucose sensor device. These needs are further defined with target specifications in the form of target values (i.e., a List

TABLE 5.1 List of Target Specifications for Point-of-Care Glucose Sensors

User Need → Metrics	Target Values	Units
Correlate with laboratory standard method	Not lower then 0.97	Correlation factor
Lot variation of consumables low	Not higher than 2%	% variation
Comply with recommendations from healthcare organizations	Comply with at least WHO guidelines	Intern. organizations
Leading to correct decision for diabetic care	95 of 100 correct	Incidence
Possible to calibrate with standards	Yes	Yes/no
No false negative	Yes	Yes/no
Low false positive	<3%	Incidence
Precision high	>95%	% precision
Accuracy high	>96%	% accuracy
Handheld	Yes	Handheld/table
Independent of sampling place	Yes	Yes/no
Limited interference of triglyceride	<5%	% interference
Limited interference of bilirubin	<5%	% interference
Limited hematocrit interference	<5%	% interference
Limited interference of ascorbate	<5%	% interference
Short sampling time	1 min	Shortest min
Short procedure for home use	3 min	Shortest min
Connected to network	Yes	Yes/no
Convenient operation temperature	15–35°C	°C
Cover a clinical relevant glucose conc. range	0.05–10 g glucose/L	G glucose/L
Small blood sample volume	50 μL	Smallest volume
Convenient instrument size	L 80 – 120 mm, H 8–20 mm	Length/height aspects
Patient/doctor convenience	92 of 100	Satisfaction degree
Low price per sample	<0.5 EUR	Acceptable price
Should be GLP/GCP adapted	Yes	Yes/no
User-friendly software	High	High/medium/low
Support provided in short time	<1 day	Days
Consumables delivered within short time	<1 day	Days

of Target Specifications as we described in Chapter 4). The values presented in the list are mainly based on literature data. As evident from the list, the targets are both performance-related with specification ranges given and qualitative with basic requirements of the design. The target specifications delimit the design possibilities as described below.

5.3 DESIGN OF BLOOD GLUCOSE SENSORS

5.3.1 Generation of Sensor Concepts

Thus, the target values in the specification are used to direct the generation of design solutions. Figure 5.3 suggests six design alternatives on a conceptual level. The figure is an example of the Permutation Chart we described in Chapter 4.

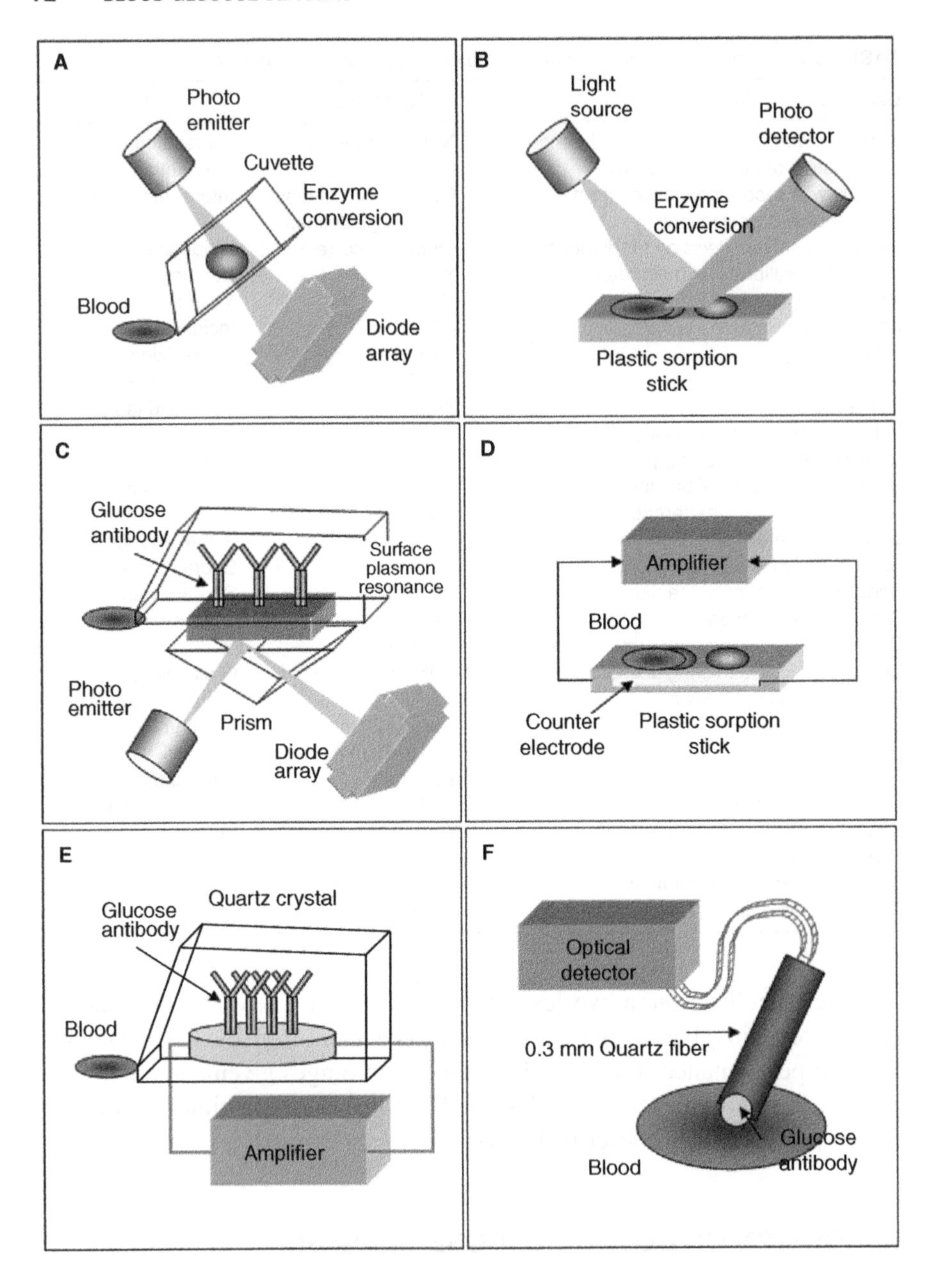

Figure 5.3 *Six alternative concepts generated for blood glucose sensors from target specifications (a Permutation Chart).*

The first alternative A is similar to the HemoCue sensor. A cuvette device is used to sample the patient's blood. The cuvette delimits the sample volume. Enzymes and reagents are contained in the cuvette and start to act on the blood cells once filled into the cuvette. A photoemitter operating at a defined wavelength irradiates the cuvette and a diode array detector measures the generated color change caused by the reaction.

Alternative B performs the enzyme conversion of the blood glucose directly on a plastic stick to which sample is absorbed. The absorption force delimits the sample volume. A color change is caused by the enzyme conversion by using dyed coreactants. This color change becomes visible on the stick. A light source irradiates the stick and the reflected light is recorded. The intensity of the reflected beam is proportional to the degree of color change on the stick. This device is similar to the Bayer device.

In alternative C, the same cuvette as in A is used for sampling and delimiting blood sample. The cuvette encapsulates a gold chip covered with a glucose-specific antibody. The chip exploits the surface plasmon resonance (SPR) effect. An optical beam impinges on the back of the chip/cuvette through an optical prism. The angle of incidence is chosen to create the SPR effect. The reflected signal becomes proportional to the glucose affinity bound to the chip. By that recording the blood glucose concentration is assessed.

Alternative D shows the same stick as in B, but in this case with electrochemical generation of the signal. Again, the advantage of sorption of a defined volume of blood sample is utilized. The glucose oxidase/ferrocene system, as shown in Figure 5.2, is a possible bioreaction to use on the stick. The reoxidation of ferrocene allows recycling of the analyte reaction product. The recorded electrochemical signal is amplified electronically in the device.

In alternative E, the SPR detection, used in alternative C, is replaced by a quartz crystal microbalance sensor. Again, a glucose-specific antibody is used to capture the blood glucose. The frequency change of the crystal is recorded and used as a response signal that is correlated with the glucose concentration. The delimitation of the blood sample volume is here more critical than in the other alternatives since volume variations may displace the frequency shift.

In the alternative F, a different approach is taken by using an optical microfiber as a small needle for blood sampling. The tip of the fiber has a small antibody spot able to carry out affinity capture of the glucose molecule in the sample, similar to alternative C. The light is coupled through the fiber to the antibody tip and reflected back in the fiber.

Once these conceptual alternatives are generated, they are screened against the target specification values listed in Table 5.1. By that the strengths of the alternative concepts are evaluated. Which of the concept can reach the thresholds of the specifications? Of those that can, how do they compete? Especially, the first question will be answered in the Concept Screening Matrix that has the purpose of making a rather coarse screening without demanding experiments and testing program, but by using a few tests and mainly literature data or other sources of information. The target specifications are converted in the matrix to selection criteria that are assessed for all of the six concepts. They are structured slightly differently to make the assessment easier. Only three levels of assessments are estimated (−, 0, +).

In Table 5.2, the result of the concept screening is presented in the Concept Screening Matrix. As is obvious, most of the values for this screening do not require any elaborate testing measurement on the concepts, but are assessed from simple tests or "common sense" judgments. In some cases, information was available in published reports.

TABLE 5.2 Concept Screening Matrix for Selection Criteria

	Concepts					
Selection Criteria	Concept A	Concept B	Concept C	Concept D	Concept E	Concept F
Functionality						
Lightweight	−	+	−	+	+	+
Size test unit	−	+	−	+	+	+
Correct decision	+	0	0	0	0	0
Convenience						
For patient	−	+	0	+	0	+
For nurse	+	+	+	+	+	−
For doctor	+	0	0	0	0	0
Time for testing	+	+	+	+	+	+
Response time	0	+	+	+	+	+
Preparation time	0	0	0	0	0	+
Availability of disposables	+	+	+	+	+	+
Regulatory acceptance	+	0	0	0	0	0
Ergonomics						
Test unit (handheld)	−	+	0	+	+	+
Disposables	+	+	+	+	+	+
Sampling time	0	0	0	0	0	+
Durability						
Storage of disposables	+	+	+	+	+	+
Lifetime test unit	0	0	0	0	0	0
Storage with sample	0	0	0	0	0	0
Performance						
Accuracy	+	−	0	−	0	0
Precision	+	−	0	−	0	0
Interference	+	+	+	+	+	+
Repeatability	+	−	0	−	0	−
Correlation	+	−	0	−	0	−
Cost						
Test unit	−	+	−	+	+	+
Cuvette/sticks	+	+	−	+	−	−
Sum +'s	14	13	7	13	11	13
Sum 0's	5	4	13	7	12	7
Sum −'s	5	7	4	4	1	3
Net score Rank	9	9	3	9	10	10

In the Concept Scoring Matrix of Table 5.3, the two concepts that are highest ranked are further scrutinized (Concepts E and F that both had the same ranks). In the Concept Scoring Matrix, the same selection criteria are assessed again but at additional levels (8 levels, marked 1–8 where 8 has the highest impact) and with weights for separate selection criteria. The weights are set based on the judgment of the design team. This introduces a subjective bias from the team that needs to be handled wisely. In this example, values between 10 and 90 have been used. Obviously, these weights directly influence the outcome of the total scoring and it is therefore advisable to compare the outcome for different setting of weights. One approach is to have two or three independent assessment groups for setting the scoring weights.

TABLE 5.3 Concept Scoring Matrix for the Two Sensor Concepts Selected in the Screening

			Concept E		Concept F	
Selection Criteria	Weight		Rating	Weighted score	Rating	Weighted score
Functionality	90					
Lightweight		30	7	210	8	240
Size test unit		30	7	210	8	240
Correct decision		30	7	210	6	180
Convenience	66					
For patient		10	6	60	6	60
For nurse		4	6	24	6	24
For doctor		2	6	12	4	8
Time for testing		10	5	50	7	70
Response time		10	8	80	7	70
Preparation time		10	5	50	5	50
Availability of disposables		10	2	20	8	80
Regulatory acceptance		10	4	40	4	40
Ergonomics	10					
Test unit (handheld)		5	5	25	5	25
Disposables		5	4	20	8	40
Sampling time		5	5	25	7	35
Durability	30					
Storage of disposables		10	4	40	4	40
Lifetime test unit		10	4	40	4	40
Storage with sample		10	4	40	4	40
Performance	60					
Accuracy		8	8	64	6	48
Precision		7	8	56	6	42
Interference		15	6	90	5	75
Repeatability		15	7	105	3	45
Correlation		15	4	60	4	60
Cost	40					
Test unit		20	5	100	5	100
Cuvette/sticks		20	3	60	2	40
Total score				1691		1692
Rank				2		2

With this setting, scoring matrix shows little difference between the two alternatives with the applied weights. Thus, none of the concepts can at this stage be ruled out. This calls for a further refinement of the concepts in order to fine-tune the design structure, understand more about the pros and cons of the concepts so that a better testing situation can be created. At this stage, a Hubka–Eder model becomes very useful.

5.4 DESCRIPTION OF THE SYSTEMS INVOLVED IN THE DESIGN CONCEPTS FOR GLUCOSE BLOOD SENSORS

A generalized transformation process (TrP) for a blood glucose sensor according to alternatives A, B, or C is shown in the Hubka–Eder map in Figure 5.4 together with systems involved in the design concept. Remember that the Hubka–Eder mapping is an analysis of the functionality of the design for achieving the transformation and not yet an attempt to configure the physical design structure (cf. Chapter 4). The consecutive flow of subtransformations of the blood sample and its treatment includes a sequence of separated steps. First, the blood sample volume is delimited in order to fix the amount of analyte molecules (i.e., glucose). Then, the blood cells are disrupted to free intracellular glucose into the extracellular space. After that, the glucose

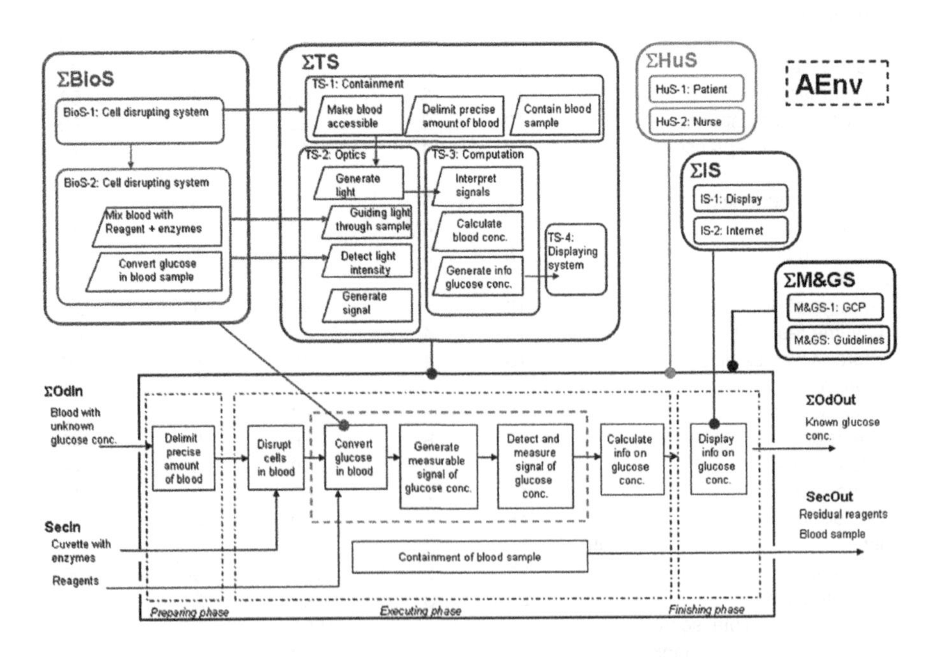

Figure 5.4 Hubka–Eder map of a glucose sensor design. Adapted from Ref. [13].

is enzymatically converted to molecules suitable for detection. This is followed by the generation of a signal from these molecules, which reflects the number of glucose molecules. Subsequently, this signal is measured and the glucose concentration in the sampled volume is calculated. Finally, the calculation is displayed for the user. Note the strict function description of the subtransformation steps. The order of the steps could be changed, but actually not very much if the purpose of the design should be met. Note also that an additional subtransformation step, containment of the blood sample, is placed separate. This is because this step may be realized in different ways.

The various systems that provide the functions for effectuating the transformation are discussed in the following sections.

5.4.1 Biological Systems

The necessary biological and biochemical systems (ΣBioS) for carrying out the TrP are enzymes and the red blood cells themselves. Contacting of blood with active reagents, either by a transport into the cells or by disruption of the cells depends on the cellular properties (cell wall, cell membrane, and diffusion conditions). The result of this is a mixture of cell components, including the glucose molecules, and the reagents. The glucose analytes may be entrapped by cellular components (e.g., membrane residues, nondisrupted organelles). This emphasizes that the blood cell itself takes an active part in influencing the transformation and requires thorough consideration.

The reagents are here thought to be either enzymes or glucose-specific antibodies. Figure 5.2 has given a few common enzymes used for converting glucose in analytical assays. However, from a functional perspective also other reaction routes justify considerations, for example, the classical colorimetric dinitrosalicylic acid (DNS) method.

The enzymatic conversions are multistep. Thus, performance, kinetics, and specificity of the acting enzymes are key design issues. The option of using recombinant enzymes and protein-engineered enzymes with more favorable functionality for the glucose assaying is an extendable design challenge.

5.4.2 Technical Systems

The technical systems (ΣTS) to be used for carrying out the TrP are a combination of sensor and actuator functions and mechanical functions.

A key subsystem is the sampling of the blood. The requirements of this system are to make the blood accessible, if possible, delimit the sample volume, protect the sample from adverse reactions or transformations, and allow detection within the device or through subsequent transformations.

In the HemoCue sensor (Figure 5.1), this has been solved by a capillary plastic cuvette also containing reagents. It has the advantage of defining the sample volume provided the cuvette itself is reproducible. The MediSense sensor and some of the other sensors rely on the surface tension of a blood droplet and its sorption onto a dry stick containing the reagents.

The generation of signal for measurement is either electrical or optical (Figure 5.3). By electrochemical reactions, electrons are generated proportionally to the amount of glucose analyte molecules. Alternatively, optical signals are generated by typical spectrometry principles, either a colored compound is produced that can allow transformation of an incident signal or light (e.g., fluorescence) is generated through a reaction. The setting for these reactions, either electrochemical or optical, is a result of the chosen ΣBioS.

The signal generated has to be detected. Optical detections such as photometers or diodes may provide that. Electrical detection via electrodes and electrical amplification circuits provides that functional requirement.

The role of the ΣTS is also to prove the technical means for the display of the result visually to the user, for example, by electronic or digital displayers. To transform the recorded signals, in a format that is suitable for the user, is the task for the information systems (ΣIS).

The function of calibrating the sensor is crucial. Calibration cuvettes/sticks with fixed absorption at the correct wavelength are seem as handy and reliable solutions. If the stability from the manufactured device is sufficiently high, at least for single use or use during shorter periods, that may be an alternative.

5.4.3 Information Systems

The main visual information function of the glucose sensors is the display of the glucose concentration in the blood sample. The value is created by transfer of generated signal data from the ΣTS based on computational routines. The calibration curve and conversion constant(s) could be more or less automatic and be updated by using a calibration procedure with calibration factors. The display can be designed differently with sound indicators, flashes, or other information functions. The role of the technical functions in the display of information to the patient/nurse/doctor is key to the accomplishment of a successful design solution and therefore this requires very careful consideration.

5.4.4 Management and Goal Systems

Under this system entity could be included functions for transfer of data to doctors and clinical management systems at hospitals.

The function of operating the device should be defined according to the Good Clinical Practice and WHO guidelines with manuals and other routines.

Procedures for servicing the sensor and distribution of disposables/reagents for the sensor must be established and could also be included as part of the management and goal system (ΣM&GS).

5.4.5 Human Systems

The aspects of the human users include patients, technicians, nurses, and physicians. The frequent use of the sensors in diabetic homecare situations emphasizes easy handling and robustness of the design. Low-cost production of the device seems strongly justified. Also, the volume of units, both the sensor itself and the disposables, makes this realistic. These issues are addressed in the functions and their effects on the design as the subsystem functions of the Human Systems (ΣHuS).

5.4.6 Active Environment

There are several unanticipated influences from an active environment. The biological variation of blood samples in the population is a significant factor to consider. The storage of more or less sensitive enzymes and other reagents where climate conditions may affect the shelf life may vary considerably. The climate variation may also influence the performance of the sensor. Sensitivity to humidity and temperature effects on the reaction rates of the analytical conversions need considerations.

By systematically mapping these systems and subsystems of the design, a deeper understanding of the prerequisites of the design structure is reached.

5.4.7 Interactions Between the Systems and Functions of the Design

The Functions Interaction Matrix (FIM) is a tool for systematically comparing the functions of the systems that carry out the transformation process. The procedure is to investigate how the separate systems and subsystems interact with each other. The importance or strength of interaction is ranked and by that the designers identify the most critical parts of the design concepts.

Figure 5.5 shows an FIM for essential interactions between the above-described systems. In a complete sensor design, the matrix in this example need to be zoomed-in to reveal more details of the design at subfunctional levels.

At the level of details shown in Figure 5.5, the matrix reveals the most critical interactions (indicated with 4 or 5) that indicate where a zoom-in is required for further investigation.

Main Systems		ΣBioS					ΣTS				ΣIS			ΣM&GS			ΣHuS		
	Subsystems	Cell disruption	Enz./Antibody activity	Active reagents	Stability enzymes Antib.	Blood stability	Containment	Mixing	Detection	Data tretment	Displaying	Calibration	Calculation	GLP compatibility	Network adapted	Support/supply	Patient	Nurse	Maintenance
Σ BioS	Cell disruption		4	5	1	1	1	5	4	1	1	4	4	4	1	1	1	1	1
	Enzyme/antibody activity	4		5	4	4	1	3	5	5	1	5	5	4	1	3	1	1	1
	Active reagents	5	4		4	5	1	4	5	5	1	5	5	4	1	3	1	1	1
	Stability enzymes antibodies	1	5	3		1	1	4	3	3	1	4	4	5	1	4	4	4	2
	Blood stability	3	2	4	1		1	4	4	2	1	4	4	4	1	2	1	1	3
Σ TS	Containment	5	4	5	5	5		5	5	3	1	2	2	3	1	4	5	5	4
	Mixing	5	5	5	1	5	1		5	4	1	4	3	3	1	2	3	3	3
	Detection	2	3	3	1	4	3	2		3	1	3	3	3	1	2	4	4	3
	Data treatment	1	1	1	1	1	1	1	1		4	4	4	4	1	2	4	4	1
Σ IS	Displaying	1	1	1	1	1	1	1	1	1		1	1	5	5	5	5	5	4
	Calibration	1	1	1	1	1	1	1	1	4	2		5	5	5	3	4	4	3
	Calculation	1	1	1	1	1	1	1	1	5	5	5		5	3	2	4	4	3
Σ M&GS	GLP compatibility	4	4	4	4	4	4	4	4	4	4	4	4		4	4	5	5	4
	Network adapted	1	1	1	1	1	1	1	1	1	5	1	1	4		4	5	5	4
	Support/supply	1	5	5	1	1	1	1	1	1	1	1	1	5	5		5	5	5
Σ HuS	Patient	2	2	2	2	5	4	4	1	2	1	5	3	5	3	4		5	3
	Nurse	4	4	1	1	1	3	3	3	3	1	5	4	5	5	4	5		4
	Maintenace	4	4	5	5	5	2	2	4	2	2	2	2	5	5	5	5	5	

Figure 5.5 *Functions Interaction Matrix for a blood glucose sensor: 5 is a very strong interaction, 4 a strong interaction, 3 an intermediate interaction, 2 a weak interaction, and 1 a very weak or nonexisting interaction.*

For example, the matrix indicates, as we would expect, that enzymes or antibodies of the biological systems have a key role in executing the analytical function of the glucose sensor. What the matrix also highlights, which maybe is not that obvious, is the high ranking of the effects of the technical systems on the biological systems. This reverse interaction is often neglected at the early stage of the design of a new sensor device.

Quite often, the alternative concepts are accepted or abandoned too uncritically before these are ranked and the functional effects benefits or drawbacks are analyzed thoroughly. Here, the FIM becomes a tool for reconsideration and systematic comparison of the conceptual strengths. Therefore, the FIM analysis tends to retard the design work at this critical point and force the designers to repeatedly and iteratively go through the generated concepts.

However, by again looking at the generated concepts in Figure 5.3 the designers will be able to more easily choose the best physical design structure.

5.4.8 Anatomical Blueprints from the Functions Interaction Matrix Analysis

The FIM analysis should pave the way for a concrete layout (blueprint) of preliminary designs. The anatomical blueprints in Figure 5.6 shall show how different biological and mechatronic components could fit and be configured into the design boundaries and preferences gained by the preceding design work. The design options, and the pros and cons, noted in the scoring and interaction matrices are used to configure the anatomic blueprints.

Figure 5.6 shows two anatomical blueprints, one for concept E and one for concept F. The components are here collected from a general search of possible components and should be considered very thoroughly preferably based on prototyping and testing. This phase of the design work will

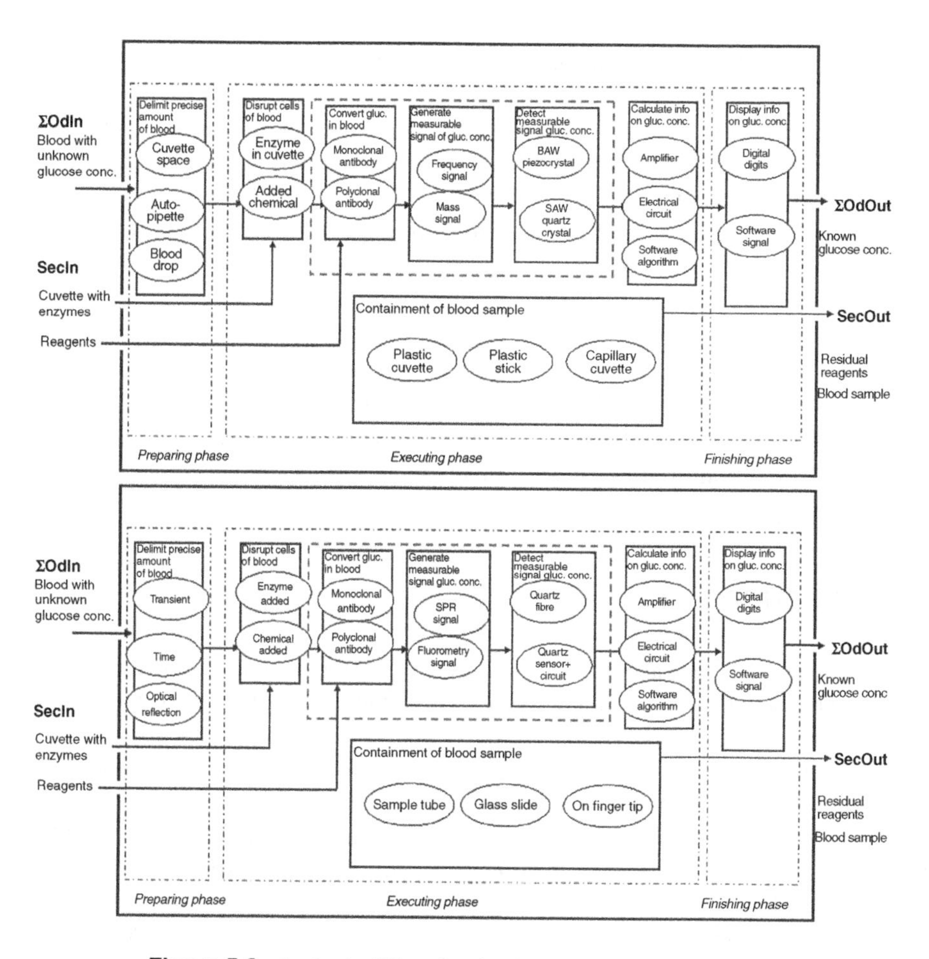

Figure 5.6 *Anatomical blueprints for the design concepts E and F.*

consequently be a key activity (using a substantial portion of the product development budget).

Here, the components are based on standard components from established vendors of antibodies, enzymes, optical devices, and microprocessors. The key components for the functions are shown in the graphs.

5.5 CONCLUSIONS

Blood glucose sensors are already well-established biotech products on the market. Thus, this design case might be considered superfluous from a product development perspective. However, this design case has had the advantage of presenting a quite transparent showcase serving as a basic illustration of the mechatronic design methodology and how it should be used in a biotechnology application.

The steps of the systematic conceptual design approach in the chapter have therefore been kept rather short without going deeper into all actual sensor design structures. Instead, it intends to open up a continuing discussion of how possibly glucose sensors could be further developed with partly different and new design concepts.

Indirectly, the chapter also serves the purpose of retrospectively reviewing the outcome of design development with a nonsystematic approach.

It is perhaps noticeable that the existing commercial products are all fairly similar in their design. At this stage, it is difficult to know if this has been caused by that other design concepts were tried and abandoned or if the commercial vendors simply copied concepts from competitors.

Thus, the chapter has tried to show that a conceptual design approach easily generates many more design solutions than appears to have been created for the existing market.

REFERENCES

1. Solnica, B., Naskalski, J.W., Sieradzki, J. (2003) Analytical performance of glucometers used for routine glucose self-monitoring of diabetic patients. *Clin. Chim. Acta* 331, 29–35.
2. Solnica, B., Naskalski J.W. (2005) Quality control of self-monitored glucose in clinical practice. *Scand. J. Clin. Lab. Invest. Suppl.* 240, 80–85.
3. Chaiken, J., Finney, W., Knudson, P.E., Weinstock, R.S., Khan, M., Rebecca, J.B., Hagrman, D., Hagrman, P., Zhao, Y., Peterson, C.M., Peterson, K. (2005) Effect of hemoglobin concentration variation on the accuracy and precision of glucose analysis using tissue modulated, noninvasive, *in vivo* Raman spectroscopy of human blood: a small clinical study. *J. Biomed. Opt.* 10, 031111–031112.

4. Buhling, K.J., Henrich, W., Kjos, S.L., Siebert, G., Elizabeth, S.E., Dreweck, C., Stein, U., Dudenhausen, J.W. (2003) Comparison of point-of-care-testing glucose meters with standard laboratory measurement of the 50 g-glucose-challenge test (GCT) during pregnancy. *Clin. Biochem.* 36, 333–337.
5. Ashworth, L., Gibb, I., Alberti, K.G.M.M. (1992) HemoCue: evaluation of a portable photometric system for determining glucose in whole blood. *Clin. Chem.* 38, 1479–1482.
6. Bellini, C., Serral G., Risso, D., Massimo, M.M., Bonioli, E. (2007) Reliability assessment of glucose measurement by HemoCue analyser in a neonatal intensive care unit. *Clin. Chem. Lab. Med.* 45, 1549–1554.
7. Hoedemaekers, C.W.E., Klein Gunnewiek, J.M.T, Prinsen, M.A., Willems, J.L., Van der Hoeven, J.G. (2008) Accuracy of bedside glucose measurement from three glucometers in critically ill patients. *Crit. Care Med.* 36, 3062–3066.
8. Hoej-Hansen, T., Pedersen-Bjergaard, U., Thorsteinsson, B. (2005) Reproducibility and reliability of hypoglycaemic episodes recorded with Continuous Glucose Monitoring System (CGMS) in daily life. *Diabetic Med.* 22, 858–862.
9. Kristiansen, G.B.B., Christensen, N.G., Thue, G., Sandberg, S. (2005) Between-lot variation in external quality assessment of glucose: clinical importance and effect on participant performance evaluation. *Clin. Chem.* 51, 1632–1636.
10. Hannestad, U., Lundblad, A. (1997) Accurate and precise isotope dilution mass spectrometry method for determining glucose in whole blood. *Clin. Chem.* 43, 794–800.
11. Püntmann, I., Wosniok, W., Haeckel, R. (2003) Comparison of several point-of-care testing (POCT) glucometers with an established laboratory procedure for the diagnosis of type 2 diabetes using the discordance rate: a new statistical approach. *Clin. Chem. Lab. Med.* 41, 809–820.
12. World Health, Organization. (2006). *Definition and Diagnosis of Diabetes Mellitus and Intermediate Hyperglycemia*, WHO Press, Switzerland.
13. Derelöv, M., Detterfelt, J., Björkman, M., Mandenius, C.F. (2008) Engineering design methodology for bio-mechatronic products. *Biotechnol. Prog.* 24, 232–244.

6

Surface Plasmon Resonance Biosensor Devices

In this chapter, design of surface plasmon resonance (SPR) biosensors is discussed. First, the theoretical background of SPR is given (Section 6.1), then the design requirements are analyzed and described from user aspects (Section 6.2), and then the biomechatronic design methodology is applied to the design (Section 6.3). Finally, the design of two of the most critical parts of the SPR is elaborated on (i) the fluidics of the sensor and (ii) the immobilization of biomolecules on the sensor chip, in finer details (Section 6.4). This design serves as a concrete illustration of how the methodology should be used in practice.

6.1 INTRODUCTION

One of the most successful biosensor achievements so far are the surface plasmon resonance sensors.

The principle of SPR, known since 1950s [1–3], was not commercialized for biological sensing purposes until the late 1980s when the first Biacore™ instrument was launched on the market [4]. This had been preceded by scientific investigations in which the design of surface-sensitive interactions on a metal film grafted on an optical prism for SPR detection was done [5].

Biomechatronic Design in Biotechnology: A Methodology for Development of Biotechnological Products, First Edition. Carl-Fredrik Mandenius and Mats Björkman.

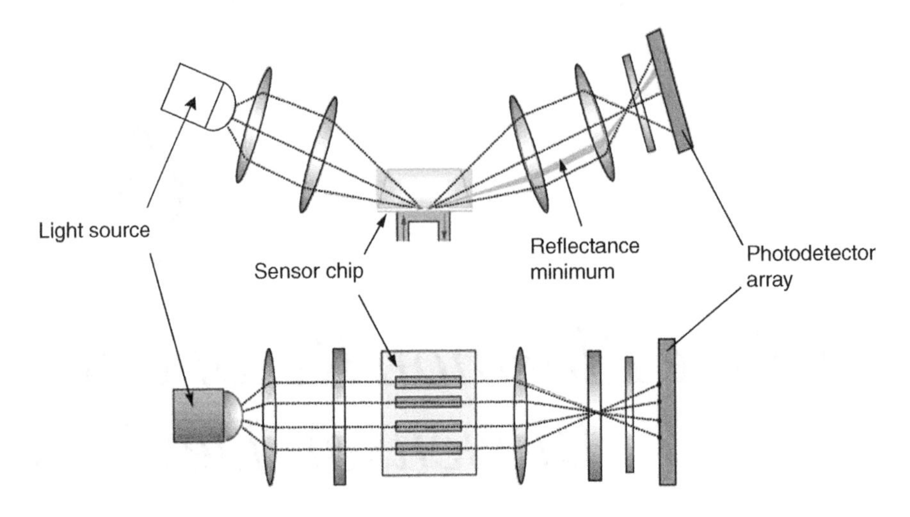

Figure 6.1 *The surface plasmon resonance principle in a typical setup. Reproduced with permission from [4].*

The optical principle of SPR is displayed in Figure 6.1. An optical beam, at a selected wavelength, impinges on a quartz prism where it reflects on a metal film covering the prism's upper surface. At a certain incident angle specific for the material, the energy propagates in parallel with the film. The energy corresponds to the quantum required for excitation of the metal's π-electrons (the so-called surface plasmon state). This quantum depends on the properties of the metal and refraction index of the surrounding media. Biomolecules covering the film influence the refractive index value. This results in a change of intensity of the reflected beam from the prism. By measuring this intensity, the molecular state of the surface is revealed. Further development of the SPR concept made it possible to quantity the mass of biomolecules on the surface with high sensitivity. Later, this was turned into a very sensitive biosensor for monitoring of antigen–antibody interactions [6,7].

Basically, the figure shows the configuration of an SPR setup as it appears in modern SPR instruments. The lower part of the figure shows a top view of a sensor chip with four parallel channels where the analytes are passed through and where interaction takes place.

The SPR chip carries the antibodies that specifically recognize the analyte (e.g., an antigen). The setup is attractive because of its small size, the possibility to monitor interactions in real time, and to run samples over different recognizing antibodies. The same chip can be used for scouting different analytes or the analyte properties for the same biomolecules. This basic principle has allowed the construction of compact systems with disposable biochips, for the user either to graft with their own choice of antibodies (or any other interacting biomolecules for a particular purpose) or to buy preimmobilized chips [4].

SPR has been used by many experimenters for a variety of investigations ranging over applications such as antibody epitope mapping, kinetic evaluation, food and clinical testing of toxicants, drug interactions, and so on [8,9].

Presently, approximately 10 vendors of SPR instruments provide the market with alternative devices of varying design. Figure 6.2 shows two

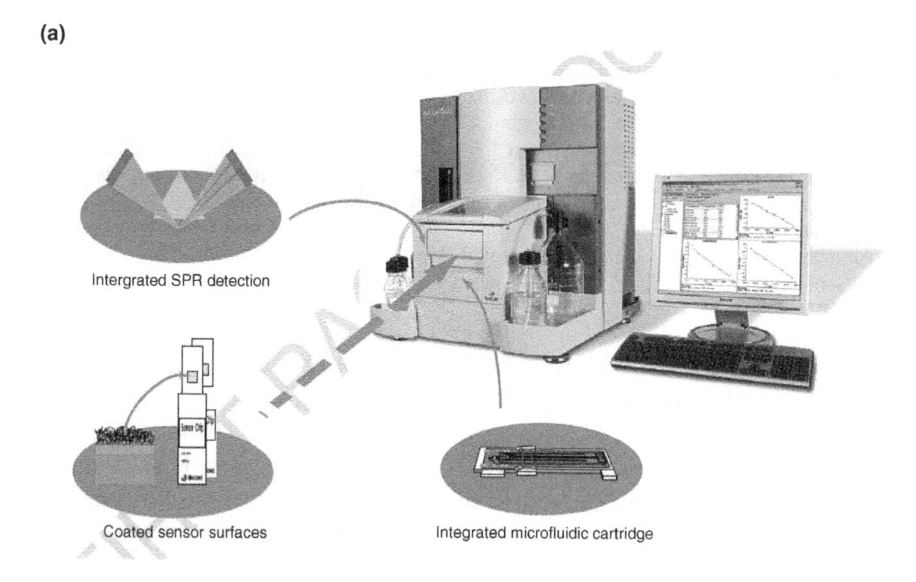

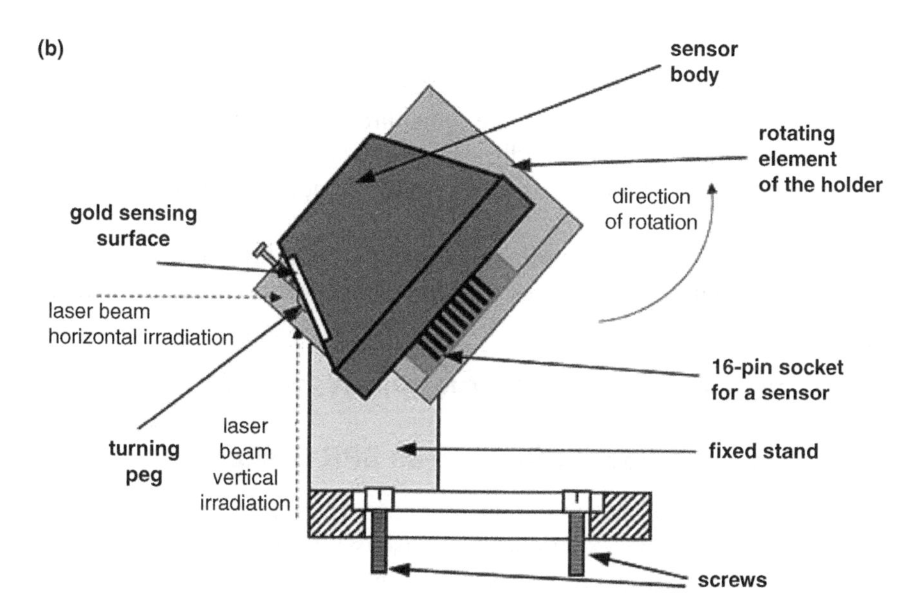

Figure 6.2 *(a) Commercial SPR system with desktop instrument, microfluidics, and biochip (Biacore, GE Healthcare). (b) Another commercial SPR system with integrated prism and chip (SPREETA, Texas Instruments). Reproduced with permissions from [4] and [10].*

TABLE 6.1 Commercial SPR Instruments: A Comparison

System	Technical Configuration	Company/Vendor
Biacore™ system	Kretschmann prism/gold chip	GE Healthcare (S)
SPREETA	Integrated sensor-prism chip	Texas Instruments (USA)
Reichert SPR system	Kretschmann prism/gold chip	Reichert Analyticals (USA)
Bio-Suplar™	Kretschmann prism/gold chip. Flow cell system	Analytical μ-Systems (D)
SPR-20		DKK-TOA Corporation (JP)
AutoLAB ESPRIT™	Double channel instrument with electrochemical detection as well	Eco Chemie (NL)
SPRiLab™, Genochip™	Compact system, SPR imaging	Genoptics (F)
SPR Imager™ II	Rotating prism	GWC Technologies
IBISiSPR™	Imaging SPR, Kretschmann configuration, scanning angle	IBIS Technology (NL)
SpectraBIO 2000		K-Mac (KR)
Plasmonic™	Compact integrated chip	Jandratek (D)
SPR Sensordisk		XanTec (D)

extremes of these: the successful Biacore system from GE Healthcare [4] and the small and compact SPREETA™ system developed by Texas Instruments [11].

The Biacore system is an expensive benchtop instrument with an advanced sensor chip construction, automated fluid handling in microfluidic system, a cartridge and a multipurpose software program for evaluation of signals.

The SPREETA system is a compact unit with the sensor chip integrated into the prism, light source, and detector [6,12,13,14]. A software program filters and presents the signals from the detector. The user has to set up all pumping and fluidic system himself. The idea is that it should be used a limited number of times (50–100), by that having the same throughput as the Biacore sensor chip. Consequently, the price per disposable unit is comparatively low, but the customers must do a lot on their own or hire the necessary competence for doing it.

Other SPR vendors offer similar instruments. Table 6.1 lists a selection of these together with commentaries on the technical differences.

6.2 DESIGN REQUIREMENTS ON SPR SYSTEMS

6.2.1 Needs and Specifications of an SPR Design

Table 6.2 compiles the most important user needs for SPR instruments. The list is based on experimental experiences and discussions with several SPR users.

The user needs are expressed or stated in terms of metrics that can be allotted a target specification value. The units are connected to the metrics and these target values.

TABLE 6.2 User Needs of a SPR Instrument and Realistic Target Specifications

User Need → Metrics	Target Values	Units
High stability	200–400	Number of runs per sensor surface
High measurement sensitivity	1,0	ng/mm sensor surface
Short analysis response time	15	s
Low detection limit	1–3	ng/mm
High precision for analysis of crude biological samples	2	%
High accuracy for analysis of crude biological samples	3	%
Compensation of nonspecific analyte binding	10–20	% of total signal
Fast regeneration time of surface	1–3	min
Insensitive to sample media variation	High	High/low
High coupling yield possible	>200	ng/mm sensor surface
Small sample volume	>100	μL
High-throughput of samples	20–60	Samples/h
Convenient operation temperature	20–35	°C
Biological operation pH	2–9	pH
Convenient instrument size	1–6	dm^3
Low price per sample	<1	EUR
Should be QC/QA adapted	Yes	Yes/no
User-friendly software	Yes	Yes/no
Powerful evaluation of signals	Yes	Yes/no
Support provided in short time	12	h
Consumables delivered within short time	24	h

Naturally, the metrics vary significantly depending on type of needs. Often a range is given for the metrics so as to match a particular need related to an analytical application the SPR analysis is intended for. These needs require a deeper consideration and should be based on an enquiry where statistical measures are also included [15].

6.3 MECHATRONIC DESIGN APPROACH OF SPR SYSTEMS

6.3.1 Generation of Design Alternatives

The target specifications restrict the possibilities of the design to certain boundaries. Still, a set of design structures can be generated where the functions necessary for accomplishing the needs and specifications are integrated in the structure.

Some functions are generic for the design purpose and will appear in all the generated design structures. In the SPR systems, the generic functions that always

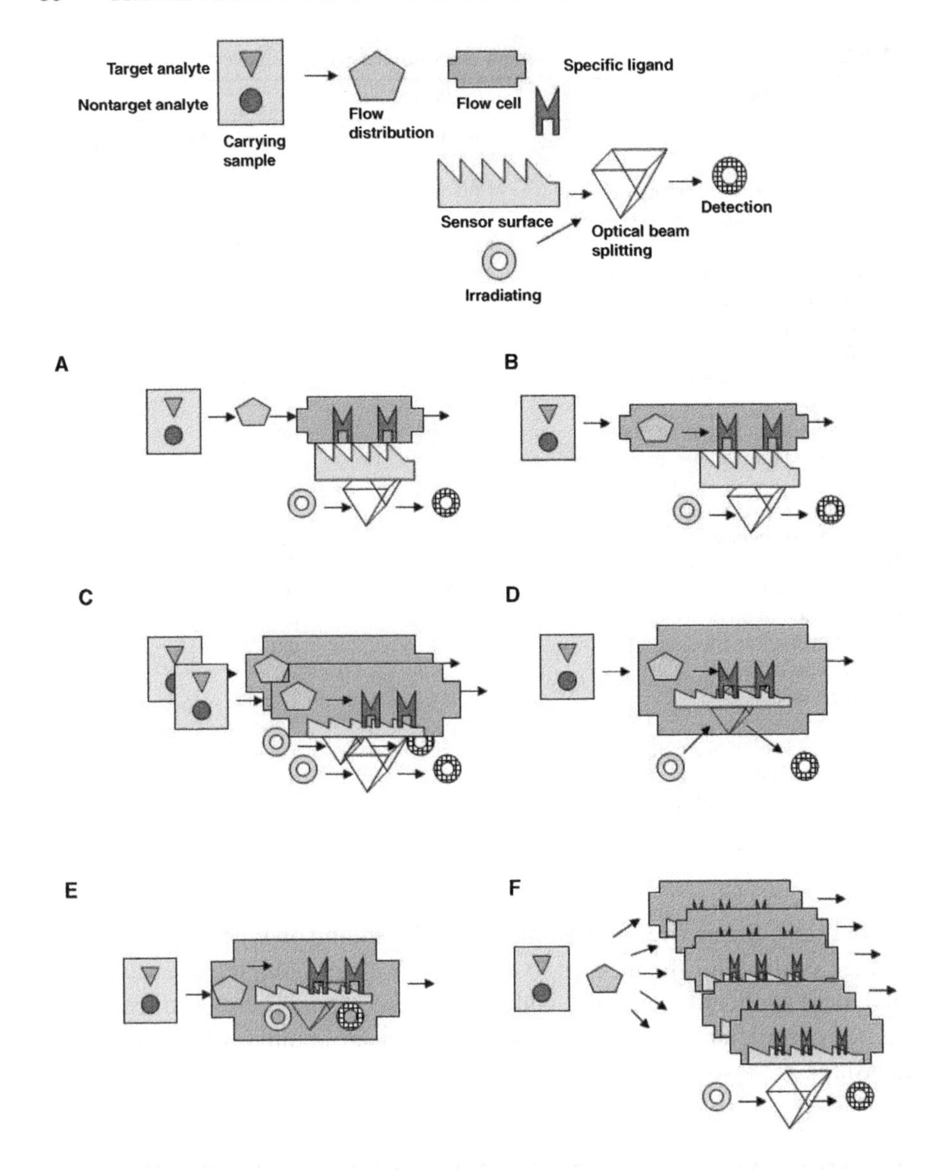

Figure 6.3 A Concept Generation Chart for the total system of the SPR design. The Permutation Chart shows here six possible configurations of the design.

reappear in the structures are the functions of the sensor surface, the optical beam splitting for SPR (Kretschmann prism), the containment of the system, the flow distribution, the irradiation of the surface, a carrier medium of the analytes, and the detection. Figure 6.3 shows the Concept Generation Chart for SPR designs.

In the upper part of the figure is the Basic Concept Component Chart that shows the generic functions together with the target analytes and the nontarget analytes. These functions are coarsely related to each other. This is done without locking the configuration with the intention to make the structures available for generating alternatives. The lower part of the figure (the Permutation Chart) shows six design alternatives that are generated from the basic concept components.

Alternative A shows a configuration where the surface function has been placed on the prism function. The flow cell system is placed close to the surface and the ligand function is placed inside the flow cell. The flow distribution is performed prior to and separate from the surface and flow cell. The irradiation and detection functions are placed before and after the prism. It is felt that this design alternative is suitable for a relative large-size system where standard components can be assembled, for example, on an optical bench, and later be encased in an instrument cabinet.

In alternative B, the A configuration remains almost unaltered but instead of separating the flow distribution and flow cell containments, they are integrated here. This opens up for a more elaborate design of a flow cell. It would give some opportunity to make the system smaller and may suggest that different construction materials such as polymeric or glass materials can be used to mend a more unique device unit.

In alternative C, the surface function has been placed inside the flow cell integrated compartment. This is a radical change. It means that the SPR surface physics shall take place in the containment unit. Also, it keeps the surface closer to the ligand function. As in B, the flow distribution is inside the flow cell. This configuration could allow a further downsizing of the essential units of the designed system while still keeping the prism, irradiation, and detection functions apart. The D configuration seems to approach the first Biacore system.

Alternative D places the prism function inside the flow cell. The other functions remain unchanged. More units inside the flow cell probably require space inside the compartment. But it is also a challenge for further miniaturization of the separate units. Here, need for a novel design can be felt.

In alternative E, the D alternative is taken further by also placing the irradiation and detection functions inside the flow cell. This means a compact design solution. It envisages a design similar to the SPREETA system shown in Figure 6.2b.

In the alternative F, the C configuration has been multiplied. This could benefit from again having the flow distribution placed separated from the flow cell, at least a part of it. The number of repeated flow cell/surface units can vary (five is depicted in the figure). For example, a Biacore 1000™ instrument has 4 channels and a Biacore A100™ has 20 channels.

The six alternatives shown in the figure are a starting point. More alternatives can be generated. However, it is clear that a deeper analysis of the configuration is required. This is preferably done by modeling the design with the help of a Hubka–Eder map of the systems.

6.3.2 Hubka–Eder Mapping of the Design Alternatives

To further evaluate the generated alternatives, it is useful to apply the Hubka–Eder map format.

As we described in Section 6.1, SPR exploits the capacity of a biological molecule to selectively capture the analyte in the sample liquid. The capturing event is transduced to a detector that is able to record the extent of the capture of analyte molecules. The SPR devices use antibodies or other recognition molecules (ligands) bound to a thin silver or gold film, often in a polymer layer. The film is attached to an optical prism in such a way that light can be multiply reflected between the ligand and film, resulting in the SPR effect. This effect is mass sensitive via the refractive index change, and so the extent of any interaction between the ligand and the analyte can be detected very sensitively.

The transformation process (TrP) helps us to identify the key actions that take place in SPR analysis. Here, we have divided these actions into three phases: the preparing phase, the executing phase, and the finishing phase (lower part of Figure 6.4).

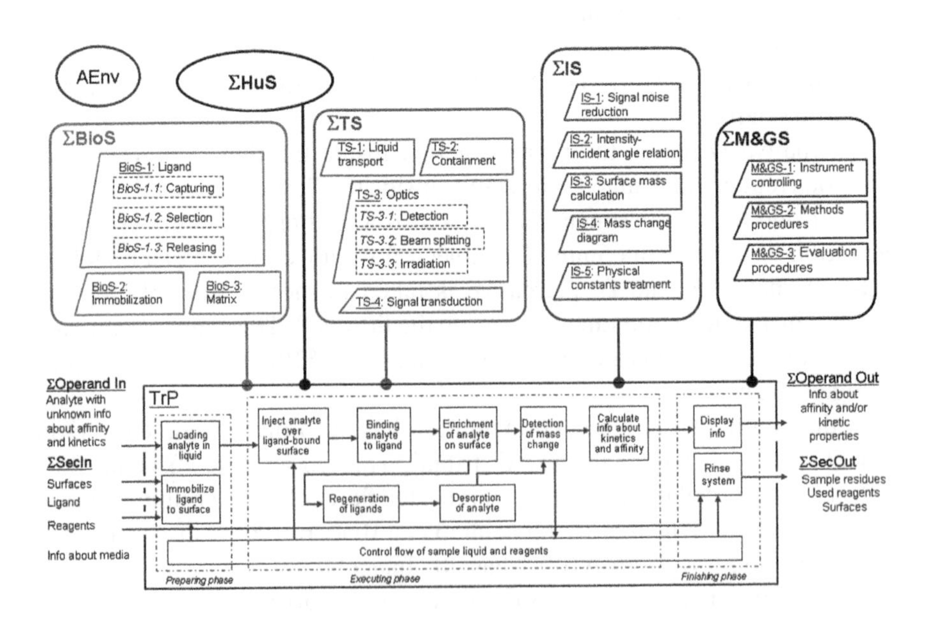

Figure 6.4 *A Hubka–Eder map for the SPR design.*

In the preparing phase, the analyte sample is loaded and the sensor surface is immobilized with ligands. When this phase is completed, the instrument can start the activity of the SPR system where the biological interaction between the analyte and the ligand is analyzed.

The executing phase do this by first injecting the analyte into the ligand-bound surface. This is followed by the binding of analyte molecules to ligand on the surface. An important subphase of the executing phase is the enrichment of the analyte molecules on the sensor surface, that is, when all available ligand sites become occupied by the analytes. These series of events are recorded in real time as mass change according to the SPR principle.

The next steps are a further evaluation of the signal and calculation of interaction kinetics. This activity can occur in real time but also be done in retrospect.

The bound analyte is removed from the surface by desorption. This has two functions: (1) to compare signals before and after the interaction and (2) to regenerate the surface for next analyte cycle. Also, these functions go through mass detection and subsequent signal evaluation.

In the finishing phase, the results of the interaction analysis are displayed.

The ingoing operand to the TrP is sample with analyte(s), and the secondary inputs are the ligand, preprepared surfaces, reagents for coupling, and liquid for transporting the samples. The outgoing operand is the analytical information, and the secondary outputs are the residual reagents, used surfaces, and sample solution.

Next is to describe the systems responsible for carrying out the TrP (upper part of Figure 6.4).

The biological systems (ΣBioS) are primarily the biological functions of the ligand molecular complex (BioS-1), and its immobilization to the sensor surface (BioS-2), and the biointerface of the sensor surface matrix (BioS-3). The ligand functions can be diversified into the capturing mechanism function (BioS-1.1), the selectivity function (BioS-1.2), and the release mechanism function (BioS-1.3).

The technical systems (ΣTS) include functions for liquid transportation (TS-1), optics (TS-2), containment of sensor surface, samples, and optical components (TS-3), and signal treatment (TS-4). Especially, the optical components are further subdivided into detection, irradiation, beam splitting, and SPR media (e.g., a metallic film).

The information systems (ΣIS) are the functions for processing the detector signal from TS-4. It concerns functions for reducing noise in order to make the signal informative (IS-1), relating angle and intensity to interpret the SPR signal dip where evanescence occurs (IS-2), to translate the signal to surface mass (IS-3), to record a time course for the mass change (IS-4), and to handle physical constants such as refractive indices for the media of the components (IS-5).

The management and goal systems (ΣM&GS) are the functional procedures for controlling the parts of the SPR instrument (M&GS-1), the method of immobilizing the surface (M&GS-2), and the evaluation of the tests carried out (M&GS-3). Also, a further subdivision can be done here.

Some of the subsystems provide materials, others energy, and others information. The interaction of operator personnel and scientists are included in both ΣHuS and active environment (AEnv).

As before, the Hubka–Eder map puts focus on the disturbances caused by the AEnv on the calculation work necessary to manage the interpretation of the kinetic results and the planning of the experiment (ΣHuS), on the requirements of the lab supply systems and the support from the vendor for instrument operation and chip fabrication and consumables (ΣTS), and on the biotechnical system, including the availability and development of suitable biological ligands and the preparation of the sample to be analyzed.

The Hubka–Eder map addresses, at an early stage, how many of the functionalities are placed outside the product (as in the Biacore instrument) and how demanding the product is, both for the user, who must provide these competences, and for the selling company, which must be prepared to provide support and consumables [16].

When each of the systems in the ΣTS domain is decomposed further, the technical operation of the biosensor instrument through the functions and subfunctions of binding of ligands to the sensor surface, sample distribution, irradiation and detection by emitter and diode array multiplier, and analysis of signals come out clearly. The relationships to materials (sample, ligands, and chips), energy (to pumps and valves), and signals (from the detector and for operating valves and pumps) can be understood similarly to a flow scheme.

Fundamental to the design work is to clearly define the main technical principle to be exploited and how the design space can be utilized in relation to that principle. Abstractly, the goal is defined as exploiting the capacity of a biological molecule (i.e., the ligand) to selectively capture or convert the analyte in proximity to a sufficiently sensitive detector.

By further decomposing the Hubka–Eder map in Figure 6.4, it becomes possible to identify additional key functions for each technical (TS-n) or biological system (BioS-n) essential for the performance and the design. Here, we focused on the decomposition to four different subsystems, each providing critical functions:

1. F_{capture}: capturing the target analyte in a complex matrix of the analyzed sample.
2. $F_{\text{mass change}}$: generating a detectable physicobiochemical change from the captured analyte.

3. $F_{\text{mass detection}}$: detecting the physiochemical change with a signal-generating device.
4. $F_{\text{analyte transport}}$: transporting the analyte sample in a time efficient way in order to carry out the above-mentioned functions.

At this early stage of the design work, the key functions for the creation of a unique new product need to be considered are $F_{\text{analyte transport}}$, $F_{\text{mass change}}$, and $F_{\text{mass detection}}$. First, the alternatives for accomplishing the mass detection function are considered and investigated. Several means for this functionality should be carefully evaluated and compared on the basis of the collection of literature data and information and, subsequently, on in-house experimental studies.

In this evaluation, critical performance criteria and other properties should be compared. These include not only typical analytical parameters such as analyte sensitivity, reproducibility, response time, and background effects but also more general criteria related to the design such as robustness, manufacturability, software processing of signals for further exploitation of the data, prospects for multisensing and miniaturization, and other integration aspects such as interfacing sample with chips and optical signals.

The importance of the design criteria for the detection means can be qualitatively weighed and thoroughly compared and ranked [16]. With the criteria listed in Table 6.3, the SPR comes out as a superior method compared to other sensory methods (e.g., ellipsometry). The next key function contributing to a unique product concept, $F_{\text{mass change}}$, is analyzed in a similar way. In order to achieve a mass change, the capturing of the target analyte must take place on a surface whose chemical structure permitted the capture to result in a mass change that was reflecting both the biological specificity and the molecular mass for the target analyte. From this means the $F_{\text{mass change}}$ could be further decomposed into subfunctions, such as *immobilize ligands*, $F_{\text{immobilize}}$, and *accumulate mass* of the analyte, $F_{\text{accumulate}}$.

The identified means for these subfunctions are mainly different metal surfaces to support the attachment of ligands, for example, gold, silver, copper, or silicon materials that can mediate the detector signal(s) and serve as a first matrix for grafting the capturing structure, various reaction chemistries for immoblizing accumulating films including thiol and silane coupling, and various organic hydrogel films, such as dextran, agarose, or poly(ethylene glycol) (PEG) in which analyte could accumulate and to which ligands could be immobilized in high yield.

These means' relations to critical performance criteria such as ability for film formation, capacity of ligand derivatization, accumulation capacity, long-term stability, sensitivity to contamination, side reactions, corrosion, and

TABLE 6.3 Evaluation Matrix for the Function $F_{detection}$. Reproduced with permission from [16]

	Criteria							
Means	Analyte Sensitivity	Manufacturability	Response Time	Referencing Options	Miniaturization Ability	Integration with Optic Systems	Sum	Rank
Surface plasmon resonance	+++	+++	+++	+++	+++	+++	18	1
Ellipsometry	++	+	+++	+++	++	++	13	2
Field effect transistor	+	++	+++	+++	++	+	12	3
Brewster-angle reflectometry	++	+	+++	+++	++	++	12	3
Surface acoustic wave	++	++	+++	+++	+	+	11	4
Bulk acoustic wave	+	++	+++	++	+	+	10	5
Photoacoustic spectroscopy	+	+	++	+	+	+	7	6

Criteria considered highly (+++), medium (++), or less favorable (+).

durability to reagents and sample solutions were analyzed. Table 6.4 compares these criteria, although, as is easily realized, the selection matrix is in reality much more complex. Table 6.4 focuses on methods that are applicable to SPR in particular.

For the biological systems, function F_{capture}, that is, *capacity of capturing target analyte*, different means are analyzed as well. The function can be attained by introducing biological ligand molecules at or very close to the place, or system, where the mass change takes place ($F_{\text{mass change}}$), and subsequently, can be detected ($F_{\text{mass detection}}$). Thus, these biological ligands become the main biological component of the biological systems ΣBioS for which the functional interconnections between the detection method, the surface chemistry, and biological capture methods become visible. Table 6.5 shows the evaluation of various ligand means versus some of the most critical properties, user's analytical purpose, to analyze a particular sample. The sample must also withstand the analyzer conditions according to the critical properties.

The fourth functional requirement, to transport the target analyte in the sample, depends on different flow conduit systems. Existing well-known means, such as peristaltic pumps, rotary valves, and silicone tubing systems were redesigned and compared in microsystem formats using various polymeric materials (e.g., silicone-based polymers, polycarbonate), hydraulic microvalves, and devices made in silicon circuits with lithographic methods. Although already listed and preliminarily ranked in the initial specification, potential analytical applications are recurrently considered on the basis of opportunities created in the functional analysis. These applications included antigen concentration determination, epitope mapping of antibodies, and kinetic evaluation of affinity interactions.

The three tables evaluate the means for each function separately and take no or little consideration of how well they fit together with each other or with other means of the final technical system. However, in order to design an "optimal solution," not only the most potential means but also the effect of the integration of the different means had to be identified and evaluated. To do so, several design solutions should be conceptually synthesized. In design science, this synthesis is commonly systemized by a morphological matrix [17]. In the matrix in Table 6.6, the functions are listed vertically and the different possible solutions horizontally. The concepts are synthesized by selection of different sets of means in the matrix. Already with this rather small morphological matrix the designers have, at least theoretically, 180 different concepts or solutions.

For obvious reasons, all combinations possible from the table should not be further analyzed. The most interesting above-mentioned alternative morphologies are compared versus general overall criteria. These included total quality

TABLE 6.4 Evaluation Matrix for Function $F_{mass\ change}$. Reproduced with permission from [16]

Means	Criteria						Sum	Rank
	Surface Stability	Reagent Stability	Corrosion	Accumulation Ability	Film Formation	Ligand Density		
Dextran/thiol gold chip	+++	+++	+++	+++	+++	+++	18	1
Agarose/thiol gold chip	+++	+++	+++	++	+++	++	16	2
PEG/thiol gold chip	+++	+++	+++	+	+	+	12	3
Dextran/silver chip	+	++	+	+++	++	+++	12	3
Dextran/silane silicon chip	+	+	+++	++	++	++	11	4

Criteria considered highly (+++), medium (++), or less favorable (+).

TABLE 6.5 Evaluation Matrix for the Ligand Capturing Function ($F_{capture}$). Reproduced with permission from [16]

	Critical Properties of the Means						
Means	Thermostability	Site Specificity	pH Dependence	Variability	Access	Sum	Rank
Antibody (Ab)	++	+++	+++	+++	+++	14	1
Nucleotide (Nt)	+++	+++	+	+++	+++	13	2
Site-specific protein	++	+++	+++	++	++	12	3
Glycoprotein	++	++	+++	++	+	10	4
Lectin	++	++	+++	++	+	10	4
Whole cell	+	+	+	++	+	6	5

Criteria considered highly (+++), medium (++), or less favorable (+).

of the biosensor, the time required for the development work, the time to market, the evaluation of various cost aspects (including R&D costs, manufacturing costs, sales and support costs), an evaluation of the market possibility/opportunities for the biosensor, and requirement of support systems of consumables (i.e., biochips, reagents, and ligands). Here, the SPR system with the stable dextran and thiol-coupled gold chips for antibody-based applications was ranked as the most favorable product design alternative, while other morphologies are considered less favorable or even unrealistic.

6.4 DETAILED DESIGN OF CRITICAL SPR SUBSYSTEMS

Two of the subsystems are critical for accomplishing a good design solution: the sensor surface and the fluidic system. In this section, we discuss how the detailed design of these two could be carried out.

TABLE 6.6 Morphological Matrix. Reproduced with permission from [16]

Function	Means[1]					
Detection of mass change	SPR	Bulk acoustic wave	Surface acoustic wave	Photo acoustic spectrom	Brewster reflectom	Ellipso metry
Sensor surface accumulation and ligand binding	Dextran/ thiol gold chip	Agarose/ thiol gold chip	PEG/thiol gold chip	Dextran/ silver chip	Dextran/ silane silicone	
Capturing analytes by surface ligand	Antibody	Nucleotide	Site-specific protein	Glyco protein	Lectin	Whole cell

[1]The dotted lines indicate desired combinations.

6.4.1 Design of the Sensor Surface

The sensor surface function is a key to the whole design. This especially could contribute to the targets settled in the specification by providing sensitivity and other performance properties of the system. In particular, it should be able to address ligand orientation, density, stability, specificity, and protection.

Obviously a deeper analysis is needed along with a more detailed configuration. Here again the generation of alternatives is an appropriate approach.

In the Concept Generation Chart of Figure 6.5, a Basic Concept Component Chart is first established. New functions are introduced. A spacer function should distance the ligands from the surface, which may have negative short-range effects. A 3D microstructure function is introduced for allowing more ligands to be loaded onto the sensor surface. A conditioning medium function is added for creating a microenvironment feasible for the biomolecules involved. A blocking function has the role of hindering and protecting vulnerable groups of the system. A secondary ligand function, primarily intended for ligand orientation, is added as an extra option. Two coupling functions, typically bifunctional reagents are included as well.

From these basic functions, the following 12 design alternatives have been generated.

Alternative A shows a configuration where the spacer, 3D microstructure, and conditioning functions are immobilized on the sensor surface by three different couplings. The space could be, for example, an alkyl chain, the 3D microstructure a hydrogel, and the conditioning a preparation of the gel with suitable electrostatic substituents. The ligand is immobilized with another coupling. Remaining active surface residues are blocked. The blocker can be a reactive bulky organic molecule that hinders access to the other functions for analytes. Probably, it can couple with the same coupling function. The entire collection of functions should be envisaged as a tight unit possible to fabricate in large quantities.

In alternative B, the conditioning function is extended above the ligands and analytes. This may have advantages is possible to realize.

In alternative C, the 3D microstructure is assumed to have the duel function of also being a space. This could be realized with some media such as PEG polymers.

In alternative D, also the condition function is carried out by the 3D microstructure. This reduces the coupling to one reagent. This would come close to a derivatized hydrogel.

Alternative E has a more elaborate blocking. It includes blocking also of the ligand parts not engaged in the capturing. Importantly, it should not interfere with the binding mechanism.

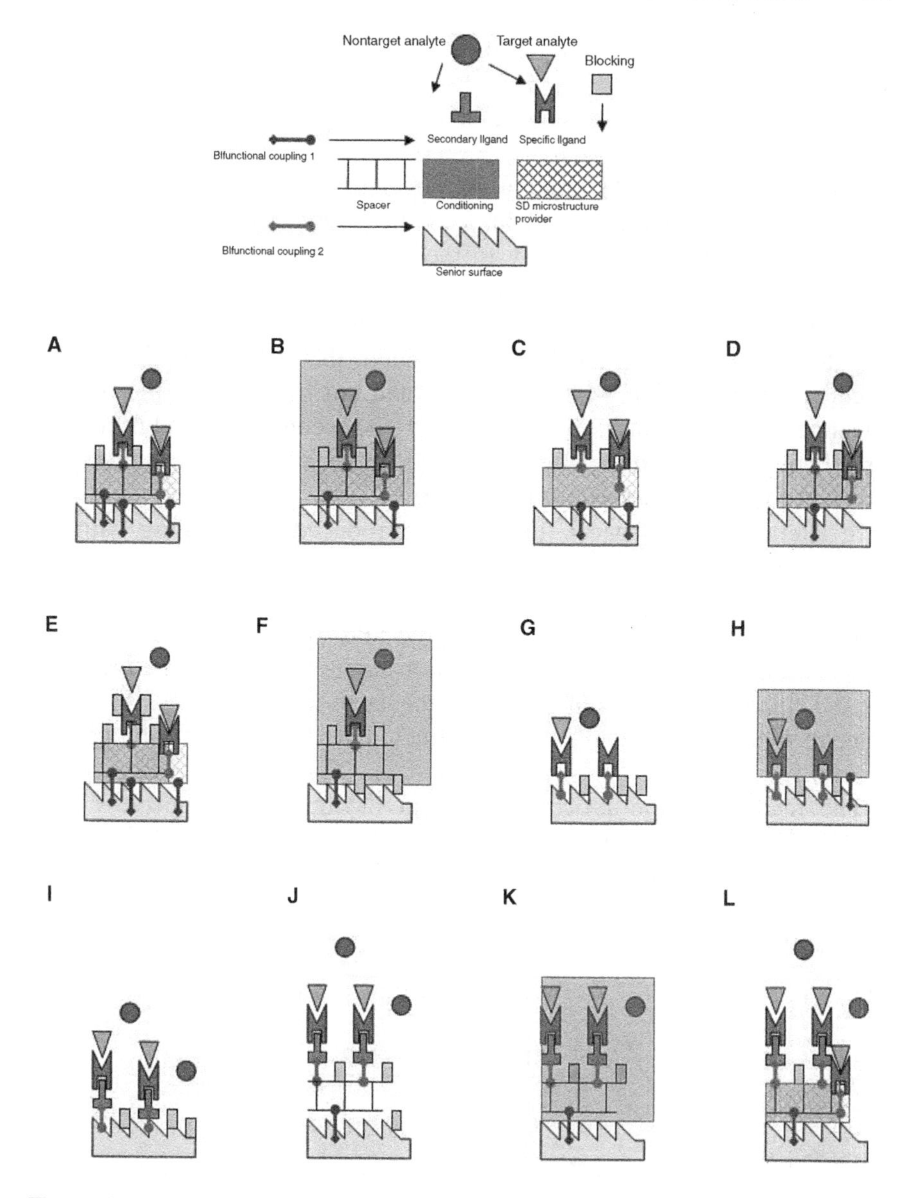

Figure 6.5 *A Concept Generation Chart for the immobilization of the chip. The Permutation Chart shows 12 possible configurations.*

Alternative F has no 3D microstructure function but relies on the space for that. It has also an extended conditioning function including the analytes.

Alternative G is a simplification of the surface where the ligand is immobilized by the bifunctional coupler 1 directly on the sensor surface.

To negate nonspecific binding, the sensor surface is densely covered by blocker molecules.

Alternative H modifies this by again introducing the conditioning medium function.

Alternatives I and J elaborate on introducing a secondary ligand that can orient the analyte binding ligand out from the sensor surface. The secondary ligand is immobilized covalently to the sensor surface while the analyte binding ligand is affinity bound to the secondary ligand. This is a mild immobilization and contributes to save vulnerable biomolecules from being lost during immobilization. It will guarantee that all bound ligands are oriented for interaction with the analyte and not hindered by the binding substrate (e.g., the spacer). In Alternative I, the secondary ligand is coupled directly to the sensor surface by the bifunctional reagent. Blocking molecules covers the nonbound areas of the surface.

In alternative J, a spacer function is introduced by coupling with the bifunctional coupling function 2 and the secondary ligand is instead immobilized to the spacer. Again, this distances the analyte binding event from the surface while still being in the range of the SPR evanescent wave.

In alternative K, the conditioning function is added, and in alternative L also the 3D microstructure is included. In the latter case, it allows the lined complex to be embedded in the 3D gel structure for increasing coupling yield further.

The immobilization procedures are in biochemical engineering well established [18]. An array of methods is available for coupling surfaces and gels with more or less advanced chemistry [4,19]. Scheme 6.1 shows some

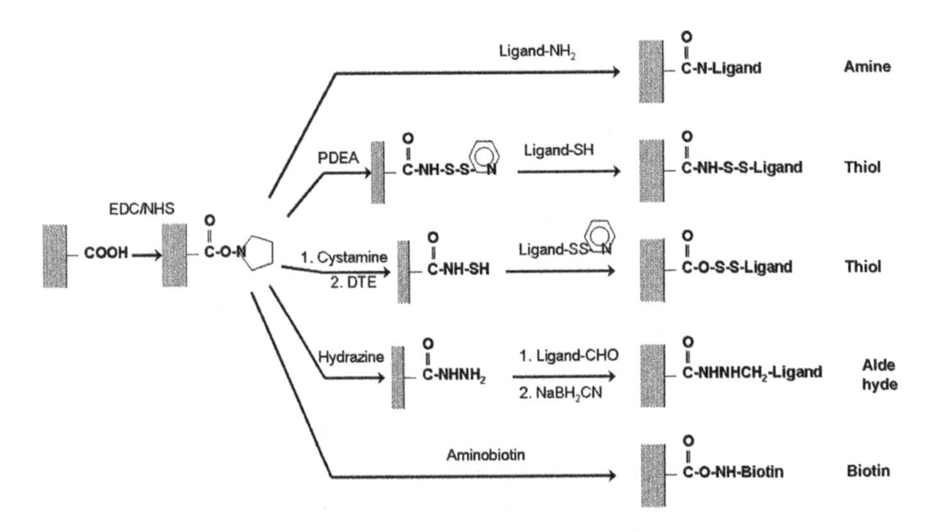

Scheme 6.1 *Alternative immobilization chemistry for SPR chips.*

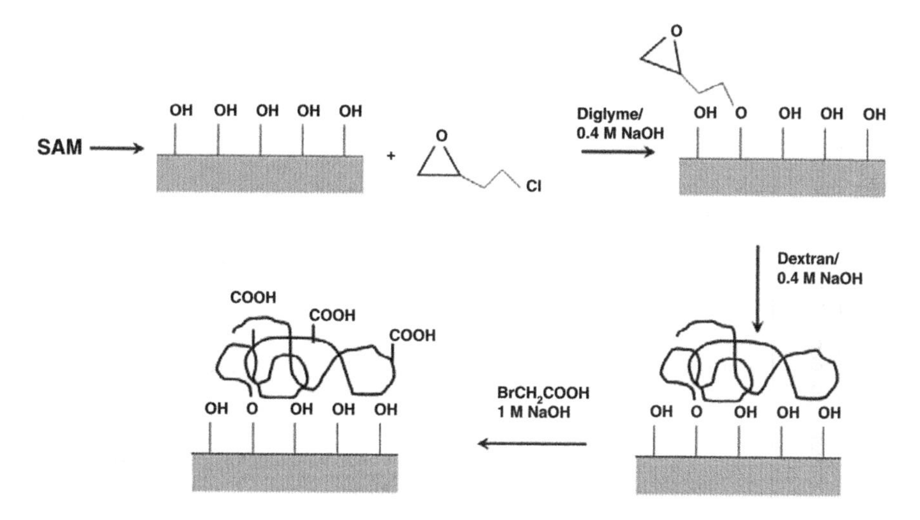

Scheme 6.2 *Alternative immobilization steps for SPR chips.*

well-known methods that are used by many. For this design case, it must be a matter of experimental evaluation to select methods and to try out how they work on the sensor surface. In Scheme 6.2, the result of such experimental work has resulted in a method for immobilizing gels, ligands on a gold surface.

These alternatives could now be assessed in a concept screening matrix based on the relevant needs previously listed (Table 6.2).

The screening is based on further studies in literature, by experiment and by preliminary prototype testing. For some assessments, interviews can be considered.

The scores given in Table 6.7 are based on our own practical experiences of SPR, suggestions by colleagues, and some supplementary enquiries. The screening results in preference for alternatives C and D.

Further elaborated studies are now required on these two alternatives. Prototypes are built that are capable of testing the primary functions the design alternative is founded on. These functions should be possible to assess with metrics that allow a ranking of the designs.

6.4.2 Design of the Fluidic System

The fluidic system refers to the transportation of liquids from sampling to exposure to the sensor surface and SPR detection. In Figure 6.4, the fluidics was represented by a flow cell and a flow distribution function. These were configured in alternative ways with the sensor surface functions.

In Figure 6.6, the functionality of the fluidic system is given in more detail. Five new functions are introduced: (1) flow distribution for spreading

TABLE 6.7 Concept Screening Matrix for Selection Criteria for Sensor Surface Concepts

	Sensor Surface Concepts											
Selection Criteria	Concept A	Concept B	Concept C	Concept D	Concept E	Concept F	Concept G	Concept H	Concept I	Concept J	Concept (k)	Concept (l)
Stability	+	+	+	+	+	+	+	+	−	−	−	−
Measurement sensitivity	0	0	0	0	0	0	0	0	+	+	+	+
Analysis response time	0	−	0	0	0	−	0	−	0	0	−	0
Low detection limit	+	+	+	+	+	−	−	−	0	0	+	+
Precision in crude solution	+	+	+	+	+	+	+	+	+	+	+	+
Accuracy in crude solution	+	+	+	+	+	+	+	+	+	+	+	+
Compensation of nonspecific analyte binding	+	+	+	+	+	+	−	+	−	−	+	+
Regeneration time of surface	+	0	+	+	+	0	+	0	+	+	0	+
Insensitive to sample media variation	+	+	+	+	+	−	+	+	−	−	+	+
High coupling yield	+	+	+	+	−	−	−	−	0	+	+	+
Small sample volume	+	+	+	+	+	+	+	+	+	+	+	+
Sample throughput	+	+	+	+	+	+	+	+	+	+	+	+
Orientation of ligands	−	−	−	−	−	−	−	−	+	+	+	+
Low fabrication price	−	0	+	+	−	+	+	+	0	−	−	−
Sum +'s	10	9	11	11	9	7	8	8	7	9	10	11
Sum 0's	0	3	2	2	2	2	2	2	4	2	1	1
Sum −'s	2	2	1	1	3	5	4	4	3	4	3	2
Net score	8	7	10	10	6	2	4	4	4	5	7	9
Rank	4	5	1	1	7	12	9	9	9	8	5	3

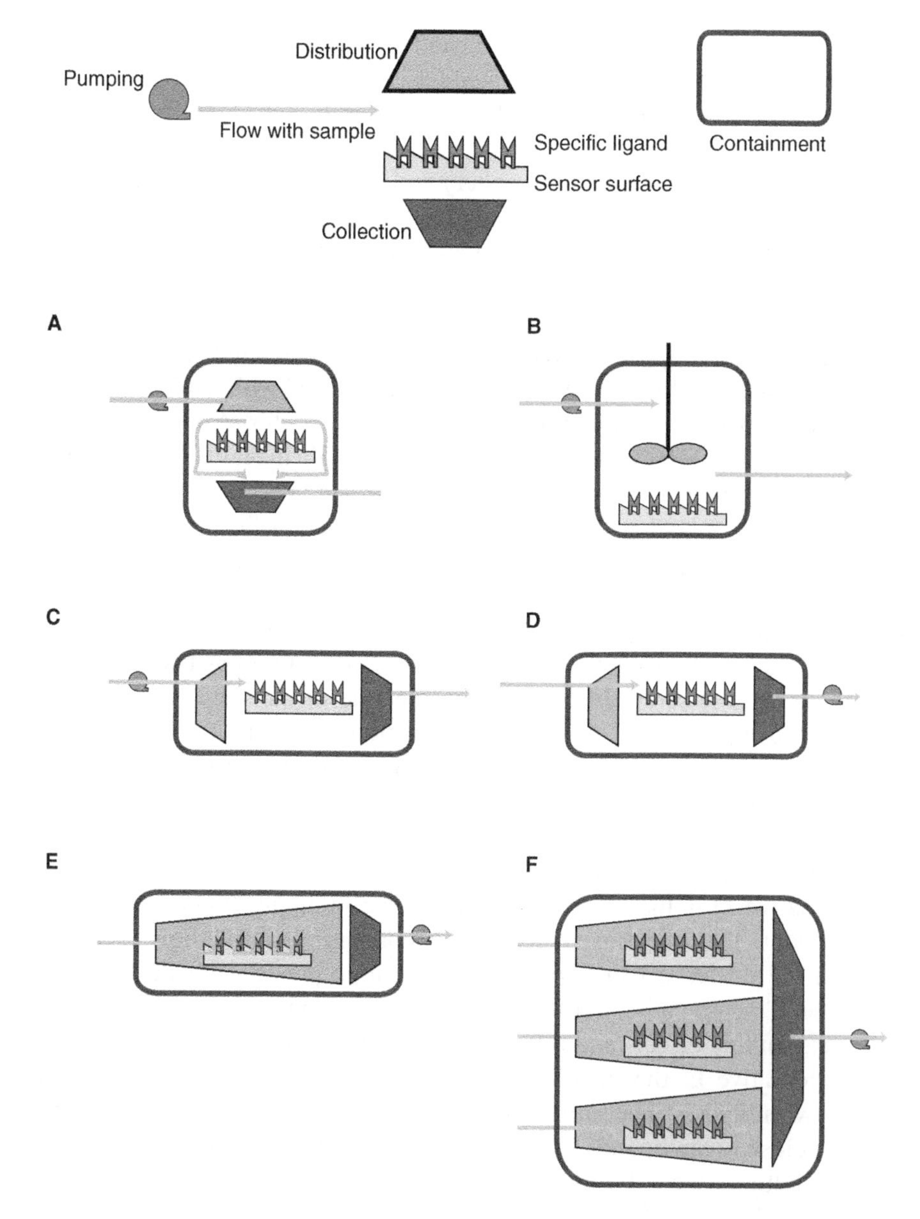

Figure 6.6 Concept Generation Chart for the fluidics system of the SPR design. The Permutation Chart shows six alternative configurations.

the analyte liquid across the sensor surface with ligands, (2) collection (and discarding) of the flow after the sensor surface, (3) a function for flowing sample liquid for transporting the liquid, (4) pumping for creating pressure to drive the liquid transportation, and (5) a containment function for casing the components.

The five functions are loosely interrelated in the top of the figure. Below, six alternative configurations are generated based on the functions.

Alternative A distributes the analyte flow over the sensor surface perpendicular from above, letting the flow spread over the ligands. Then the flow leaves the ligand surface at the periphery and becomes collected beneath the sensor surface. The functions are assembled in the containment function, for example, a compact plastic chamber. Only the pumping is placed outside the container. The flow distribution could, for example, be a helical coil starting at the center moving across the ligand surface until it reaches the periphery. It could also be a thin space between a distributor plate and the sensor surface.

In alternative B, this arrangement has been replaced with a microchamber with stirring, similar to a continuous stirred tank reactor. A microagitation function is added (represented by a propeller, but could in reality be something else causing mixing) on top of the sensor surface. The flow passing in is instantaneously mixed and distributed, thus the agitator fills the distribution function. The flow passes out from the chamber due to the pumping on new sample liquid into the containment chamber. Here, we can envisage, for example, a metallic microdevice with electrically generated vibrations.

In alternative C, the stirred tank fashion configuration has been replaced with a transversal setup. Flow enters at one side of the sensor surface and leaves at the opposite side. A distribution function is placed at the entrance and a collector at the outlet. The containment shapes the setup to resemble a tube or column into which liquid is pumped in. In this alternative, an integrated device in polymeric materials can be foreseen. It would probably be easier to miniaturize than alternative B.

Alternative D is the same as C but has the pumping function after the containment. This may have a certain advantage: the sample flow is sucked through the device causing lower hydrodynamic pressure inside the tube, while the C setup causes an increase in hydrodynamic pressure. This may influence mixing effects and diffusion of analytes to the ligands.

In alternative E, the distributor has been extended to cover the sensor surface with ligands. The sample flow is coiled across the surface, forming a channel close to the ligands. To realize this arrangement, a special design of the distributor is needed. Probably, a lithography fabrication of the geometry channel could be a solution.

Alternative F is a multiplication of the sensor surface in the same containment in order to multiply the capacity of the device. The E arrangement is the most suited for this. The distribution function will now also have the role to direct the flow to several sensor surfaces. The pump is placed after the container. The number of sensor surfaces in on container can vary.

The six alternatives are screened in a matrix as above for critical target specifications as listed in Table 6.8.

TABLE 6.8 Concept Screening Matrix for Selection Criteria for Fluidics Concepts

	Fluidic Concepts					
Selection Criteria	Concept A	Concept B	Concept C	Concept D	Concept E	Concept F
Stable flow	+	+	+	+	+	+
Measurement sensitivity	+	+	+	+	0	0
Analysis response time	0	−	−	−	+	+
Low detection limit	+	+	+	+	+	+
Precision in crude solution	+	+	+	+	+	+
Accuracy in crude solution	+	+	+	+	+	+
Compensation of nonspecific analyte binding	0	0	0	0	0	+
Regeneration time of surface	0	0	0	0	0	0
Insensitive to sample media variation	0	0	0	0	0	0
Coupling yield total	+	+	+	+	0	0
Coupling yield per area	0	0	0	0	+	+
Small sample volume	0	0	0	0	+	+
Sample throughput	0	−	−	−	+	+
Low fabrication price	+	+	+	+	+	+
Sum +'s	7	7	7	7	9	10
Sum 0's	7	5	5	5	5	4
Sum −'s	0	2	2	2	0	0
Net score	7	5	5	5	9	10
Rank	3	4	4	4	2	1

For stable flow pattern it is without rheological studies difficult to compare the concepts. It seems plausible that all have reasonably good flow pattern. Therefore, the same assessment (+) has been granted to the sex alternatives. For measurement sensitivity, it is assumed that using the whole surface as in alternatives A–D is more favorable than only using the area demarked by the channel path. Therefore alternatives A, B, C, and D have been assigned a + while alternatives D and E have been assigned zero.

For analysis response time, alternatives A, B, C, and D have been considered slower due to the larger volume in the container where transport and diffusion of analytes probably take longer time. Alternatives E and F should be faster due the smaller volumes and thereby distances, while alternative A could perhaps be miniaturized for this purpose.

Detection limit, accuracy, and precision in crude biological solutions are assumed similar in all alternatives. They have been assigned here with a + since the arrangements are all favorable.

In alternative F, compensation for the nonspecific effect is easy by using one of the channels for estimation of the effect.

Coupling yield distinguish between total and area yield. We believe the flow channel configurations are much better in enhancing the coupling than the more voluminous setups. This also favors sample volume and throughput.

The fabrication price factor is difficult to estimate. In principle, soft lithography and other small-scale fabrication techniques can be used for all

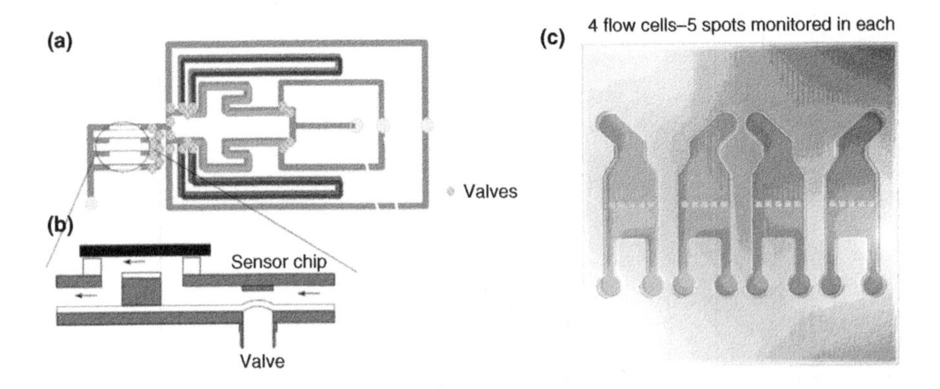

Figure 6.7 Comparing with an existing SPR flow distribution system: (a) integrated flow circuitry (Biacore 1000™), (b) design of chip flow channels (Biacore A100™), and (c) combinatory use of channels. Reproduced with permission from [4].

alternatives equally. If so, mass production of units should be easy once an assembling fabrication line is setup and automated.

The summation of the assessments ranks alternative F as the best and alternative E as the second best. Thus, this alternative should be paired with the sensor chip design according to alternatives C and D. Since these two are probably very similar from the fluidics point of view, any of these can be used with fluidics alternative F.

In Figure 6.7, we see examples of this setup principle from the Biacore 1000™ and A100™ systems. It seems that these design solutions coincide with the conclusion of the mechatronic design analysis, while a more integrated design as in SPREETA is not.

However, it must be emphasized that screening should be followed by scoring assessment that would be based on more experimental evidence.

6.5 CONCLUSIONS

This chapter has shown in detail how the mechatronic conceptual design theory is applied. It has shown the benefits of initially defining needs and target specifications, continuing with generating design alternatives based on functions and ranking these. It has also shown how the Hubka–Eder mapping further analyzes these functions and relates them to the transformation process, thereby increasing the understanding of the interactions of the subsystems of the design.

In particular, the chapter has shown how the design methodology can be useful for zooming-in critical and pivotal parts of the design, and by that, bring these down to more concrete design solutions.

Here, in this chapter, we have not intended to go through all parts of the methodology. Chapter 5 brought up other aspects of the design methodology that also can be used within this example, such as functions interaction matrices and anatomical blueprints.

REFERENCES

1. Ritchie, R.H. (1957) Plasma losses by fast electrons in thin films. *Phys. Rev.* 106, 874–881.
2. Turbader, T. (1959) Complete adsorption of light by thin metal films. *Proc. Phys. Soc. Lond.* 73, 40–44.

3. Reather, H. (1988) Surface plasmons on smooth and rough surfaces and on gratings. In: Höhler, G. (ed.), *Springer Tracts in Modern Physics*, Springer-Verlag.
4. Löfås, S. (2007) Biacore – creating the business of label-free protein-interaction analysis. In: Marks, R.S., Cullen, D.C., Karube, I., Lowe, C.R., Weetall, H.H. (eds.), *Handbook of Biosensors and Biochips*, Wiley.
5. Liedberg, B., Nylander, C., Lundström I. (1983) Surface plasmon resonance for gas detection and biosensing. *Sens. Actuators* 4, 299–304.
6. Marchesini, G.R., Koopal, K., Meulenberg, E., Haasnoot, W., Irth, H. (2007) Spreeta-based biosensor assays for endocrine disruptors. *Biosens. Bioelectron.* 22, 1908–1915.
7. Karlsson, R. (2004) SPR for molecular interaction analysis: a review of emerging application areas. *J. Mol. Recogn.* 17, 151–161.
8. Rich, R.L., Myszka, D.G. (2010) Grading the commercial optical biosensor literature. *J. Mol. Recogn.* 23, 1–64.
9. Rich, R.L., Myszka, D.G. (2007) Survey of the year 2006 commercial optical biosensor literature. *J. Mol. Recogn.* 20, 300–366.
10. Matějka, P., Hrubý, P., Volka, K. (2003) Surface plasmon resonance and Raman scattering effects studied for layers deposited on Spreeta sensors. *Anal. Bioanal. Chem.* 375, 1240–1245.
11. Soelberg, S.D., Chinowsky, T., Geiss, G., Spinelli, C.B., Stevens, R., Near, S., Kauffman, P., Yee, S., Furlong, C.E. (2005) A portable surface plasmon resonance sensor system for real-time monitoring of small to large analytes. *J. Ind. Microbiol. Biotechnol.* 32, 669–674.
12. Haasnoot, W., Marchesini, G.R., Koopal, K. (2006) Spreeta-based biosensor immunoassays to detect fraudulent adulteration in milk and milk powder. *J. AOAC Int.* 89, 849–855.
13. Spangler, B.D., Wilkinson, E.A., Murphy, J.T., Tyler, B.J. (2001) Comparison of the Spreeta (R) surface plasmon resonance sensor and a quartz crystal microbalance for detection of *Escherichia coli* heat-labile enterotoxin. *Anal. Chim. Acta* 444, 149–161.
14. Chinowsky, T.M., Growa, M.S., Johnston, K.S., Nelson, K., Edwards, T., Fua, E., Yager, P. (2007) Compact, high performance surface plasmon resonance imaging system. *Biosens. Bioelectron.* 22, 2208–2215.
15. Ulrich, K.T., Eppinger, S.D. (2007) *Product Design and Development*, 3rd edition, McGraw-Hill, New York.
16. Derelöv, M., Jonas, Detterfelt, J., Mats Björkman, M., Mandenius, C.-F. (2008) Engineering Design Methodology for Bio-Mechatronic Products. *Biotechnol. Prog.*, 24, 232–244.
17. Pahl, G., Beitz, W. (1996) *Engineering Design: A Systematic Approach.* Springer, Berlin.

18. Hermanson, G.T., Malllia, A.K., Smith, P.K. (1992) *Immobilization Affinity Techniques*, Academic Press, New York.
19. Johnsson, B., Löfås, S., Lindquist, G. (1991) Immobilization of proteins to a carboxymethyl dextran modified gold surface for biospecific interaction analysis in surface plasmon resonance. *Anal. Biochem.* 198, 268–277.

7

A Diagnostic Device for Helicobacter pylori Infection

The chapter describes the design of a clinically successful test system for diagnosis of *Helicobacter pylori* infection based on detection of emission from the bacteria in the patient's breath. First, the epidemiology and the variety of diagnostic principles for the infection are described (Section 7.1). Then, the diagnostic needs are specified with a typical biomechatronic design perspective on existing breath analysis methods (Section 7.2). This is followed by generation and selection of different design alternatives for breath testing (Section 7.3). The most highly ranked alternatives are further analyzed by Hubka–Eder mapping and compared with one existing commercial test (Section 7.4). The purpose of the chapter is to illustrate for the readers how a typical bioanalytical technology such as the *H. pylori* breath analysis can utilize the biomechatronic design methodology for generating optimal design solutions.

7.1 DIAGNOSTIC PRINCIPLE OF *HELICOBACTER* INFECTION

The infection of the upper gastrointestinal tract, stomach, and duodenum by the bacterium *H. pylori* is one of the most contagious diseases in the world. Fifty percent of the world population is infected by *H. pylori* in upper

Biomechatronic Design in Biotechnology: A Methodology for Development of Biotechnological Products, First Edition. Carl-Fredrik Mandenius and Mats Björkman.

gastrointestinal tract [1]. However, more than 80% of the infected individuals are not symptomatic, and normally it takes long to develop the disease. Chronic low-level inflammation of the stomach lining leads to duodenal and gastric ulcers and stomach cancer [2]. The *H. pylori* bacterium penetrates the mucoid lining of the stomach and cause there severe ulcers [1].

The *H. pylori* is the helical shaped Gram-negative bacterium. The bacterium produces acidic enzymatic reactions causing lesions in the gastrointestinal organs. High activities of urease, catalase, and oxidases, all involved in these reactions, are potential candidates for diagnostic assaying for identifying infected patients. *H. pylori* has also characteristic outer membrane proteins, some available for immunoassays.

Diagnostic testing of the infection is considered a challenge in clinical analysis [3]. This is partly due to that *H. pylori* is considered to lack absolute reliability as an indicator of the gastric ulcer infection [4]. Therefore, so far, it is recommended to use combinations of test methods for reliable diagnosis [5–7].

Currently applied diagnostic methods are summarized in Figure 7.1. Very reliable are microbial cultures of endoscopic biopsies from the gastrointestinal stomach. The drawback is the long culture time taking at least 1 day and sometimes difficult microscopic identification of the infecting species. Diagnosis can also include antibiotic susceptibility testing by culturing for planning therapy. Alternatively, polymerase chain reaction (PCR) methods can be used for critical genes [5].

Faster methods are serological tests [8]. For example, blood antibody testing of urease or other sufficiently specific biomarkers for *H. pylori* is possible to run within 1 h but requires analytical equipment, skilled practice, and expensive reagents. Another possibility is urine sampling followed by ELISA assaying. Also, stool samples are considered reliable. PCR analysis of specific genes of *H. pylori* has a very high identification precision but is costly and time consuming. Histological examination of biopsies, although also time consuming, has a high reliability and precision for diagnosis [6, 7].

In recent years, urea breath tests (UBTs) have become increasingly popular. The UBT exists in several variants [9–12]. Two isotopes of urea are used in the tests, the ^{13}C or the ^{14}C urea isotope, which are detected with different methods. Citrate is usually added to enhance the reaction. The false negative responses are normally low.

The UBT systems are considered cost-effective in the healthcare system and are therefore often the method of choice. To invent or refine better UBT devices or kits is consequently a prioritized product development activity.

One of the already commercially realized UBT systems is the Heliprobe™ test (Figure 7.2). The Heliprobe uses ^{14}C-labeled urea ingested by the patient in a tablet form. The tablet is dissolved in the stomach and urea is, when

Test system	Principle	Reference
Test kit Helicobacter STOOL TEST KIT 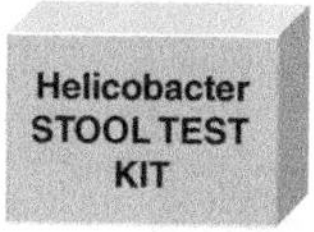	Stool tests: e.g. using IgA ELISA and/or antibody to *Helicobacter pylori* antigens	Commercial products available
Test kit 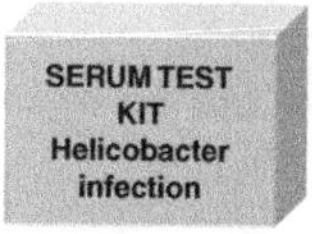 SERUM TEST KIT Helicobacter infection	Serum test: qualitative Western blot assay for detection of IgG antibodies to *Helicobacter pylori* in human serum or plasm for CagA, VacA and urease subunits	Commercial products available
HeliProbe 	UBT system based on radioactive ^{14}C labelled urea and breath care. Geiger-Müller counter is used.	www.kibion.com
Kit Urea Breath Test KIT 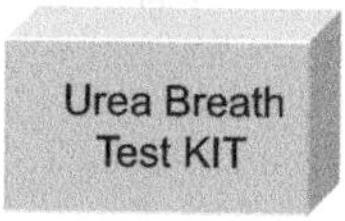	UBT ^{13}C substrate is consumed; patient breath sample tube. The breath samples are analysed by isotope ratio mass spectrometers.	Commercial products available

Figure 7.1 *Current methods for diagnosis of Hp infection (graphical illustration of alternatives): culture, serology, stool, UBT, and PCR. Reproduced with permission from Kibion.*

H. pylori is present, converted. The reaction product, CO_2, is collected from the breath of the patient by using a credit card-sized collector and detected. The Heliprobe system has been compared with several of the alternative diagnostic methods described above. These comparisons have shown good agreement with various patient groups. The Heliprobe system has exhibited high sensitivity (96.6%) and specificity (100%) compared to endoscopic methods [13, 14]. After eradication therapy, the ^{14}C-UBT showed 100% specificity in children [15]. In adults, similar results were obtained [16]. The vendor of the Heliprobe system assesses its product characteristics against alternative diagnostic methods (Table 7.1). However, room for improvement and further development is recognized.

The purpose of this chapter is to analyze the UBT systems to show how the biomechatronic design methodology can be applied to this type of

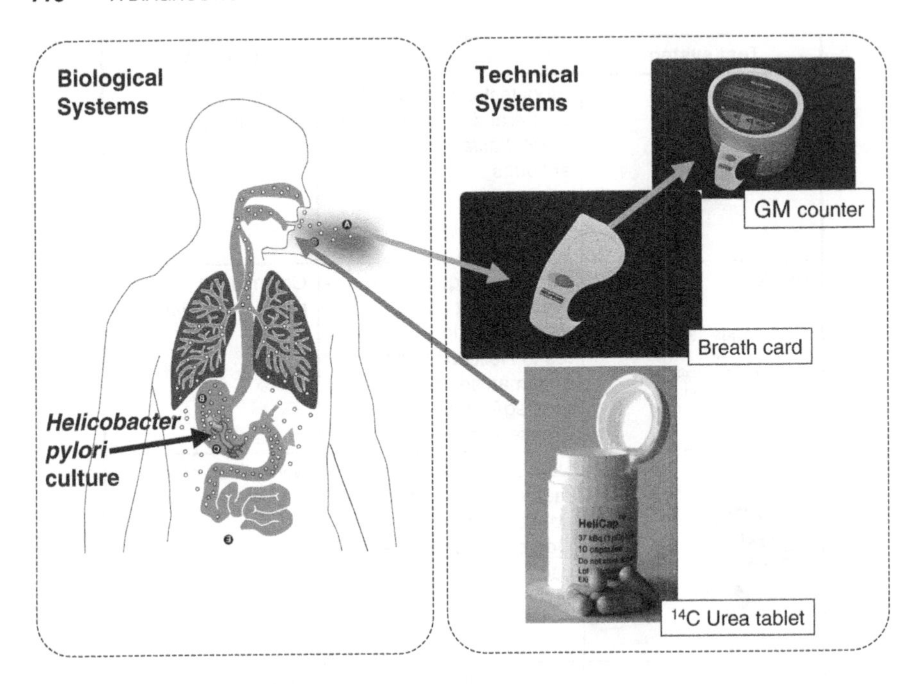

Figure 7.2 The Heliprobe UBT system (picture or drawing). The relationship between the parts of the system. Reproduced with permission from Kibion AB.

biotechnology design defined by a system consisting of four parts: a biological system (ΣBioS) confined in a human system (ΣHuS) in the form of the patient, a technical consumable for collecting sample (ΣTS), a detection device (ΣTS), a pharma vehicle transporting the reagent to the ΣHuS. The analysis is verified on the Heliprobe UBT system that is commercially performed (Kibion AB, Uppsala, Sweden). Key design questions that are brought up are as follows: Could the design have been done differently? Could it have been realized with

TABLE 7.1 Ranking of Diagnostic Alternatives

	Endoscopic Sampling	Serology	Liquid UBT	HeliCap[a] (Heliprobe)
Patient comfort and convenience	+	++	++	+++
Time	+	+++	+	+++
Preparation	+	++	++	+++
Cost	+	+++	++	+++
Sensitivity and specificity	+++	+	++	+++

Source: Adapted from Kibion AB.
[a] The tablet containing ^{14}C urea.

other gears/components and for what reasons? Could a more convenient UBT system be designed from the perspective of the users (patient, nurse, doctor, and supplier/service functions). We compare the Heliprobe with other alternative design solutions.

7.2 MECHATRONIC ANALYSIS OF UREA BREATH TEST SYSTEMS

First, we analyze the generic UBT system according to the design methodology presented in Chapter 4 to generate a collection of design alternatives [17]. The principles outlined in Chapters 3 and 4 are followed.

7.2.1 Mission and Specification for a Urea Breath Tests

The mission of the design in a UBT system is to be able to meet the present need of diagnosis of *H. pylori* infection. The mission presupposes that the test system is integrated in the diagnosis methodology applied for the *Helicobacter* infection with other tests for supporting elimination therapy of the infection. Since the UBT methods have shown high relevance for predicting the infection in comparison with other diagnostic methods, the mission has been set to establish the UBT system in a commercial instrumentation.

Table 7.2 gives the identified customer needs for such a system for healthcare use. The table specifies the needs and provides the metrics with values based on current reports on typical values.

Critical characteristics of the needs that were identified are typical analytical performance attributes such as sensitivity, specificity, analysis time, sampling time, repeatability, and correlation factors with reference or golden standards.

Another group of needs is related to the convenience of the test. Should it be handled by the patient? Should it be handled by a nurse? In what environment should it be used?

The next category of needs relates to the regulatory demands. The regulatory agencies (e.g., FDA and EMA) or others (e.g., WHO and national bodies) have set up requirements or recommendations that preferably should be met.

The manufacturability needs are also added to the list. What shall be the unit manufacturing cost? How shall the supply chain be arranged? What is an acceptable price level for the instrument per test?

These settings may provide a framework for the technical and biological possibilities of generating design alternatives.

TABLE 7.2 Customer Needs for a UBT System and Realistic Target Specifications

User Need → Metrics	Target Value	Units
High stability	1	Longevity of the ingesters, in years
High measurement sensitivity	000	Sensitivity of the detector for labeled CO_2 analytes in cpm/sample
Sensitivity	98–99	%
Selectivity	>90	%
Short analysis response time	>30	Time in minutes for developing CO_2 for achieving a sensitivity as above
Lower detection limit	$\geq 10^4$	Minimum number of bacteria detected per individual, bacteria per individual
High precision for analysis of crude biological samples. Deviation from reference method low	±2	%
Correlation with golden standards	99	% correlation with standard
High accuracy for analysis of crude biological samples		
High-throughput of samples	≥10	Maximum number of samples per hour with the same instrument setup
Convenient operation temperature	15–35	Operation range in °C where above specification is attained
Biological operation pH	2–8	Gastrointestinal pH range accepted
Small sample volume		0.0 L (Np) breath volume of 30 s exhale
Convenient instrument size	2	Instrument cage volume in dm^3 for a benchtop version
Patient/doctor convenience	High	High/low
Low price per sample	3 EUR	Sample price in EUR
Should be QC/QA adapted	Yes	Yes/no
FDA recommendation/ approval	Yes	Yes/no
User-friendly software	Priority FDA approved	Regulatory agencies accepting software
Support provided in short time	≤2	Days for exchange of detector
Consumables delivered within short time	≤2	Days for delivering spare part ingestion unit and exhaust container unit

7.2.2 Generation of UBT Design Concepts

In Figure 7.3, the basic design components are displayed in a Basic Concept Component Chart and in Figure 7.4 are displayed six design alternatives generated in a Permutation Chart. The specifications above for UBT have been guiding the permutations shown. Some of them exist as commercial systems.

Concept A is a UBT system where ^{13}C urea is ingested in a liquid cocktail, breath is collected by blowing in a straw in an absorbent, and the absorbent is injected into a mass spectrometer for quantification. Concept B is a system

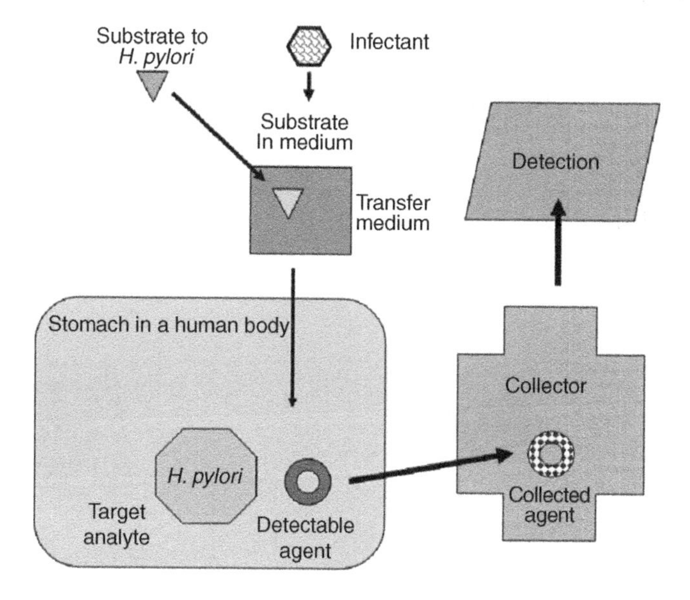

Figure 7.3 *Basic Concept Component Chart for the UBT design.*

where radioactive ^{14}C urea is used thus requiring a radioactivity counter for quantification. In this concept, the breath is collected by a solid phase (absorbent) confined in a credit card-sized container. In concept C, the ^{13}C urea is injected intravenously and sampled as in concept A, that is, by blowing in a straw placed in a liquid absorbent to be analyzed by a mass spectrometer. Concept D uses again radioactive ^{14}C urea, formulated in a tablet used in combination with a blow card and a counter. In concept E, the liquid cocktail is replaced with a tablet formulated with the same constituents as in the liquid cocktail. In concept F, the reaction product is collected by an absorbent with a dry phase reaction in a blow card and analyzed by a counter. Certainly, many more concepts can be generated with variant and permutations, especially if introducing other physiochemical principles for detection, other isotope combinations and other separation devices of the breath samples. The six concepts shown here primarily serve for elucidating the subsequent reasoning.

7.2.3 Screening and Scoring of UBT Design Concepts

In Table 7.3, the design alternatives above are screened in relation to the most important specification demands with estimated values versus assumed target metrics. Three levels are used for screening. In some instances, these were deduced from own practical experiences. More often, they are based on published literature data where testing of similar devices is presented. To that are added interviews with users of UBT systems and preliminary trials. The

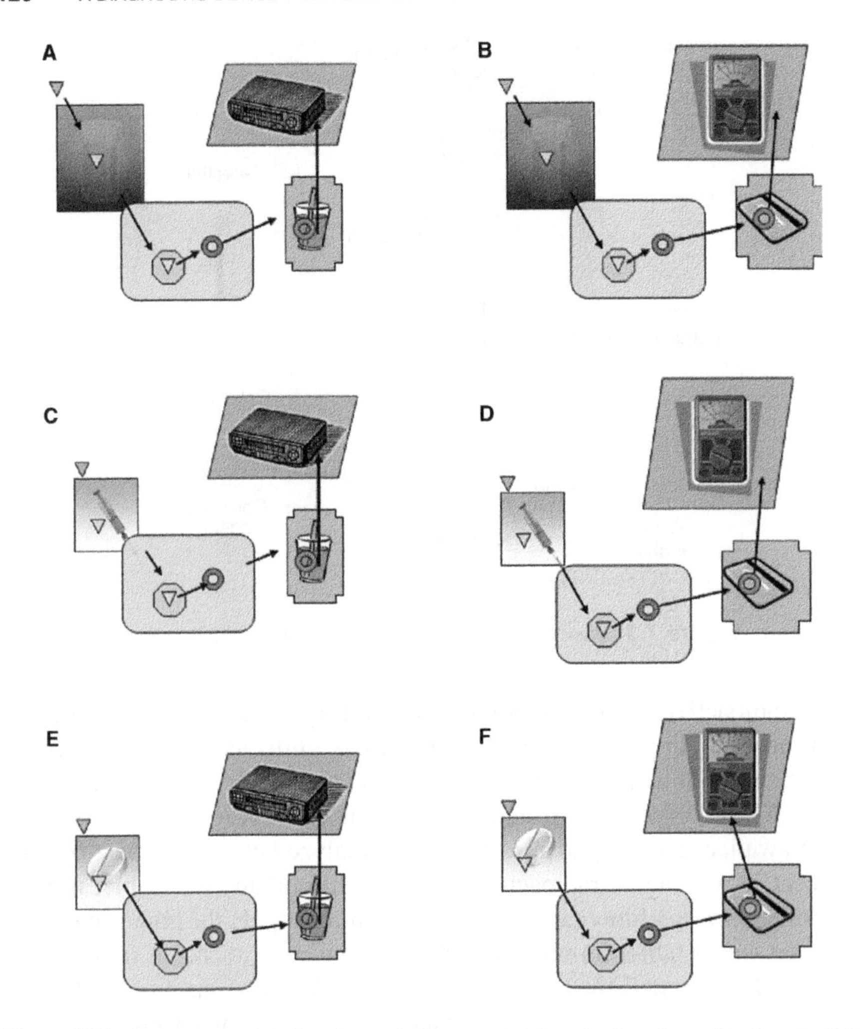

Figure 7.4 *Six conceptual design alternatives generated from the target specification of a UBT system in a Permutation Chart.*

screen summary (net score) results in a ranking of the six concepts. The table shows that concept F gained the highest score followed by B and D, while the concepts A, C, and E all had even negative scores. Thus, the former were selected for further evaluation.

In the concept scoring matrix shown in Table 7.4, these three best alternatives are scored based on further experimental testing of the UBT methods. The scoring matrix has weight for the selection criteria. The ground for the weight metrics is based on the discussions in the development team and interviews and enquiries with potential users, market experts, and production developers. The weights finally decided to be used will undoubtedly bias the

TABLE 7.3 Concept Screening Matrix for Selection Criteria

Selection Criteria	Concepts					
	Concept A	Concept B	Concept C	Concept D	Concept E	Concept F
Functionality						
Lightweight	–	0	–	0	–	+
Size detector unit	–	–	–	0	0	0
Size breath collector	–	+	–	+	–	+
Size ingesting method	–	0	+	–	0	+
Convenience						
Number of units	0	0	0	0	0	0
Time for testing	–	0	–	0	–	+
Preparation time	–	–	–	–	–	+
Spare part availability	0	0	0	0	0	0
Radioactive component	–	+	–	+	–	+
Ergonomics						
Detector unit	0	+	0	+	0	+
Ingestion method	–	–	–	–	+	+
Breath collector	–	+	–	+	–	+
Durability						
Storage of ingestion unit	0	0	0	0	+	+
Storage of unused collector	–	+	–	+	–	+
Storage of collect with sample	–	0	–	0	–	0

(*continued*)

TABLE 7.3 (*Continued*)

	Concepts					
Selection Criteria	Concept A	Concept B	Concept C	Concept D	Concept E	Concept F
Performance						
Time for testing	–	0	–	0	0	+
Selectivity	+	0	+	0	+	0
Sensitivity	+	0	+	0	+	0
Repeatability	+	0	+	0	+	0
Correlation with golden standard	+	–	–	–	–	+
Cost						
Detector unit	–	+	–	+	–	+
Collection unit	–	+	–	+	–	+
Ingestion unit	+	+	0	0	+	+
Spare parts	0	0	0	0	0	0
Sum +'s	5	8	4	7	6	16
Sum 0's	5	12	6	13	7	8
Sum –'s	14	1	14	4	11	0
Net score	–9	4	–10	3	–5	16
Rank	5	2	6	3	4	1

TABLE 7.4 Concept Scoring Matrix for Selection Criteria

		Concept B		Concept E		Concept F	
Selection Criteria	Weight	Rating	Weighted Score	Rating	Weighted Score	Rating	Weighted Score
Flexible use	20						
Doctor's office	5	5	25	4	20	6	30
Home use	10	3	30	3	30	6	60
Clinic	5	5	25	4	20	6	30
Performance	40						
Test time	10	5	50	6	60	6	60
Sensitivity	10	4	40	8	80	7	70
Selectivity	10	6	60	6	60	6	60
Repeatability	10	6	60	5	50	7	70
User convenience	25						
Easy training	5	5	25	4	20	8	40
Support	10	7	70	7	70	7	70
Test preparation	5	6	30	7	35	8	40
Healthcare integr.	5	6	30	6	30	6	30
Cost	30						
Investment	5	6	30	4	20	6	30
Consumable cost/test	10	7	70	7	70	7	70
Operation cost/test	15	6	90	6	90	7	105
Manufacturability	35	5	175	5	175	5	175
Total score			810		830		940
Rank			3		2		1

outcome of the scoring. Thus, it is of utmost importance to make careful decisions and thorough weight estimates. The rationales used in Table 7.4 are based on a thorough analysis of the competition situation for UBT systems, the assessment of the healthcare systems criteria for diagnostic methods, and the set standardization by international committees for clinical gastroenterology. As can be noted, performance weights are highest, which is based on the fact that the credibility of test results has priority in most Western healthcare systems and these countries are the prime markets for the product. The convenience and flexibility weights are also high due to the appreciated general level of new instrumentation in healthcare. Manufacturing price is also provided with a high weight due to the commercial target value of the producers of the UBT systems.

The score number is on a scale of 1–8 where 8 is the optimal performance of the established selection criteria. Each criterion has a tabulated score value. The score values are associated with the specification target metrics. Again, test programs with representative samples and individuals have been considered.

7.3 DESCRIPTION OF THE SYSTEMS INVOLVED IN THE DESIGN CONCEPTS FOR THE UREA BREATH TESTS

The transformation process (TrP) that should take place in the UBT system can be divided into three phases [18,19]. First, urea is going to be ingested and transported to the *Helicobacter* infection. Second, the conversion of urea to CO_2 by the bacteria-derived urease shall be executed. Third, the formed CO_2 should be collected and measured. The primary operands are the ^{14}C urea and a formulation. Secondary inputs are probably a disposable sampler and breath of the patient. Possibly, the patient data can also be input into the analytical system. The operand out from the TrP is the obtained analytical data. The secondary outputs are exhaust breath, used samplers, and residual formulation.

This is not the only possible configuration of the TrP. What characterizes this is that the patient and device are both included. This has been chosen for more optimally describing the effect interactions of the systems. The Hubka–Eder map in Figure 7.5 shows the transformation process, its phases, and inputs and outputs.

7.3.1 Biological Systems Involved

The biological systems involved in the design of the diagnostic instrument are (1) the bacterium, (2) the patient's organs involved in the diagnostic

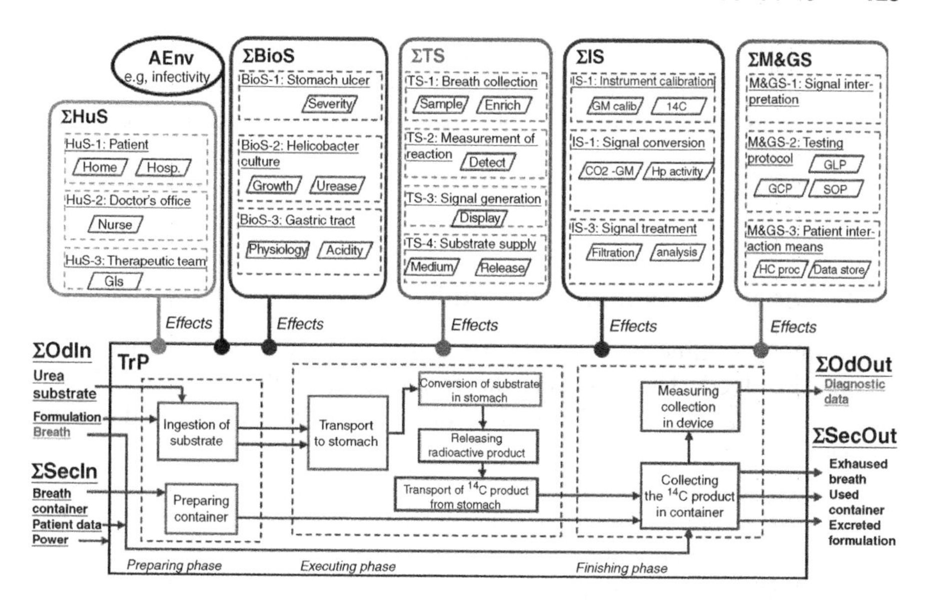

Figure 7.5 *Hubka–Eder map of a UBT system (cover Heliprobe and other test systems).*

procedure (the gastrointestinal tract, stomach, mouth, and lungs), and (3) interfering bacteria.

The target bacterium, *H. pylori* excretes urease into the stomach where the extracellular urease converts the ingested ^{14}C urea to $^{14}CO_2$ according to

$$(NH_2)_2{}^{14}CO + H_2O \rightarrow {}^{14}CO_2 + 2NH_3$$

The upper gastrointestinal tract, stomach, or duodenum harbors the *H. pylori*, where it causes the inflammation and gastric ulcers. The physiological conditions of the organs become a part of the bioanalytical system since they affect the reaction and consequently the performance of the UBT system.

The action of the bacterium in the stomach on the transformation can be translated into a number of rate-controlled conversion steps, involving the release of the enzyme and its activity due to the microenvironment in the stomach.

The action of the physiological conditions of the gastrointestinal tract for releasing the urea substrate is also part of the biological system role in the UBT system.

The adsorption of the residual reactants does also involve the gastrointestinal tract.

The biological transformation reaction can potentially be carried out by other invading bacteria in the biological system. This could be the mouth bacterial flora. Thus, additional biological systems must be considered in the design as well. Figure 7.5 shows the biological systems that may cause effects on the UBT design.

7.3.2 Technical Systems Alternatives

Based on the outline of the TrP several different technical systems (ΣTS) become involved. These are (1) the formulation of the substrate, (2) the collection of sample, and (3) the detection systems of the collected analyte molecule.

The formulation of the substrate is, in principle, a pharmaceutical design solution. It is an oral formulation, either a tablet or a liquid. The composition of the components of a tablet and its fabrication could follow typical procedures for pharmaceutical tablet preparation. If a liquid formulation is preferred, a cocktail should be considered. The formulation shall release ^{14}C urea in the stomach. It can effectuate various side reactions that facilitate uptake and pH control.

$^{14}CO_2$ is exhaled in the breath and the breath is collected for analysis. This could be a bag or an absorbent. In one of the systems considered, it is a so-called breath card that the patient blows in until it is saturated. An indicator dye shows when the absorbent is filled. Thus, the function is to collect gaseous sample with $^{14}CO_2$ in a quantitatively representative way so the subsequent detection of the CO_2 formed by urease is made.

The absorbed sample of CO_2 shall be determined. Since it is labeled with ^{14}C or possibly some other isotope, the detection function can be carried out with typical radiochemistry methods such as β-scintillators or with Geiger–Müller counters. Alternatively, mass spectrometry may be used where the mass number for the isotope is recorded.

The counts per minutes or the mass spectrometry signals become the analytical response.

The detection function is selected based on response time, sensitivity, analytical cost, measurement time, convenience for measurement, and so on. These properties are to interact with the sampling method, thus the interactions between these subsystems are critical for the design.

Furthermore, the manufacturability of the chosen technical systems should be considered. Design for manufacturing analysis will include unit cost, manufacturing time, reproducibility of sampling device (e.g., card), and reproducibility of formulation of tablet/liquid. These should be factors in the ranking of methods/solutions.

7.3.3 Information Systems (ΣIS) Required

The information systems' (ΣIS) key component is found in the technical system of the detector. The signal is displayed to the user based on software interface. The functionality of the software becomes an important part of the design where calibration of signals to the user information is defined. This may include functions for statistics and more or less customer specified parameters. If the customer shall have the opportunity to adjust the functions or if the parameters are set by the manufacturer of the UBT system, there are other alternative design choices.

7.3.4 Management and Goal Systems Required

The software can contain several important management and goal systems (ΣM&GS) functions. The clinical practice and the regulatory requirements of the test systems are probably set with the supervision of the regulators or according to other standardization bodies or committees. If these functions are embedded in the software, it is an important design decision.

Distribution of cards and tablets should also be on the functions of the test system. How these functions interact with different alternatives of the ΣTS and ΣIS may result in different priorities of the design solutions. Cost for effectuating the function is definitely a key consideration for market success of the UBT system.

The detector system is also a part of this. The device is either repaired or discarded when there is malfunction and a new detector device is provided. This affects the design.

Other service function concerns expert advice, how easy instrumental faults are diagnosed, and so on.

7.3.5 Human Systems Involved in the Testing

The human systems involved are the patient to be diagnosed, the nurse carrying out the test, the doctor receiving the diagnostic results, and possibly other member of the therapy team.

The patient's role in the diagnosis is to ingest the urea formulation and breathe into the collection device. The ways different patients do this shall be considered in the design of different alternatives of formulations and collection devices.

The nurse's role is to interact with the patient through instructions and to carry out the manual procedure of the sampling device and the detector device. The nurse shall also interact with the doctor about the diagnostic results.

The doctor's role is to receive the information from the display of the detector through the nurse. The doctor shall probably also interact with the therapy team.

The supply of tablets and cards are to be done by a service function at the doctor's office/clinic, where different systems are followed depending on the country or healthcare system.

Consequently, the analysis of the functions on the ΣHuS involved in the use of the UBT system are numerous and complex *per se*. The variation may be considerable depending in the marketing strategy. For the design issue, these consequences are extensive to take into account when evaluating the design alternatives.

7.3.6 Active Environment That Can Influence

Finally, the active environment (AEnv) plays a role in the design. Here, we decided to include unknown bacteria in the body as part of the ΣBioS. Thus, the focus of the active environment is on how conditions in the environment influence the performance of the other systems. Ambient conditions have effects on the devices.

Are there any environmental conditions that are unique for this type of test system?

Could regulatory framework change? Will the disease affect a significant part of the world population change? Will there be side effects due to other treatments of patients? These types of questions can be asked but are not always possible to reply.

In Figure 7.6, an anatomical blueprint of the Hubka–Eder map is set up. The concept alternatives in the Permutation Chart are captured in the blueprint. One of the components should be selected and the Scoring Matrix has suggested which. However, the ΣBioS, ΣTS, ΣIS, and ΣM&GS should now be scrutinized versus the anatomical alternatives. This will be further exemplified in the next chapter.

7.4 ASPECTS OF THE DESIGN FOR EFFICIENT MANUFACTURE

The cost has always been very important for the majority of products. This has been the case especially for "standard" products where price often is an *order winner*. An order winner is a product feature that is very important for getting the order from the customer. For more advanced products, performance and quality have been order winners and the cost has often been of limited importance. The price has instead been an *order qualifier*. That is, the

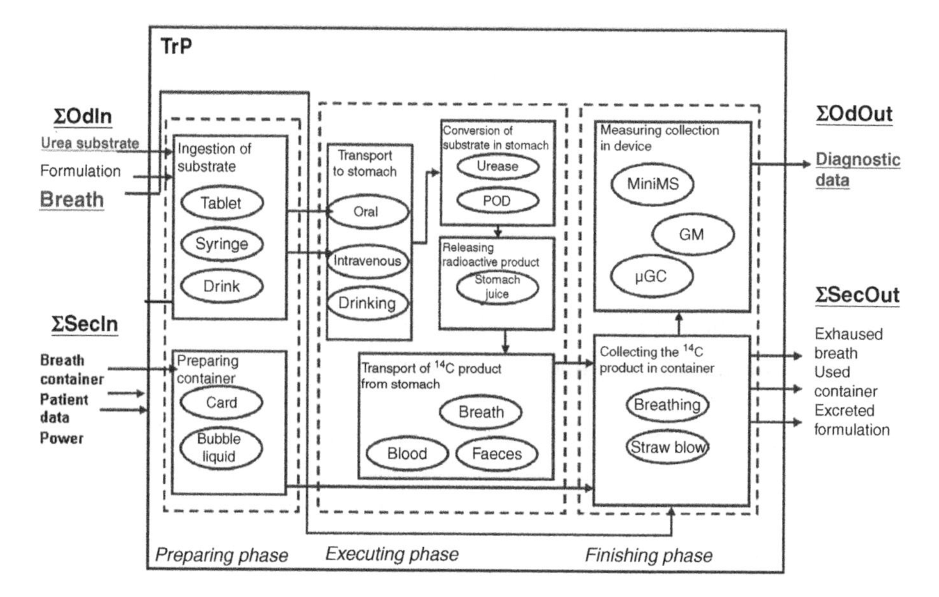

Figure 7.6 *Anatomical blueprint of the UBT design.*

customers are not price sensitive but there is an upper limit for the price that cannot be exceeded. However, today the price has become an order winner or at least an important order qualifier for more types of products where the customers traditionally not have been very price sensitive. This is the case for many of the types of biomechatronic products that this book addresses.

Diagnostic devices for *H. pylori* infection and blood glucose sensors are typical biomechatronic products where cost is very important. The product is intended for a typical biotechnical mass market and there exist a number of competitive concepts and producers. This makes the price a potential order winner, while performance and other features are more of order qualifiers that the product must fulfill.

Products that are more in the forefront of science and technology are often not so price sensitive from market point of view. Often, there are a limited number of competitors on the market. Furthermore, performance and features are considered well worth paying for. Bioartificial organ-simulating devices and manufacturing systems for stem cell manufacturing can be seen as examples of such products. However, the market may change rapidly and the market for these products may soon become price sensitive. This aspect makes it vital to consider the cost aspect in the early development and design phases of these types of products. Otherwise, there is a great risk that the original design is given expensive characteristics that are very difficult to change later.

It is also important to remember that cost and price are not the same thing. It is the market that decides the price of the product: the price is not set by the cost. The greater the positive difference between price and cost, the greater will be the profits for the selling company. However, a change in market due to, for example, new competitors may result in a rapid decrease in price. This may result in huge difficulties for a company if the costs are too high and it is difficult to reduce them quickly.

There exist two major means to achieve a cost-efficient product design. These are as follows:

- A concurrent and integrated design process.
- Use of design for manufacturing/assembly tools.

A concurrent and integrated design process is a process where the product and the manufacturing system for the product are designed concurrently and integrated by a cross-functional team with complementing competences and knowledge [17] (see also Chapter 2.7.1). The decisions that govern the main costs are often taken very early in the development and/or design process. This makes it vital that the manufacturing engineers are involved from the very beginning of the development and design of new products. It is vital that the manufacturing engineers are given a real influence decision-making power on the design of the product. The biomechatronic design methodology in this book can act as an important instrument for supporting the communication between the designers and the manufacturing engineers.

There has been a lot of work done in the field of tools for design for manufacturing, assembly and costs (DfM, DfA, and DfC). One of the most influential researchers in this field is G. Boothroyd. *Automatic Assembly*, coauthored by him with C. Poli and L. Murch and published in 1982, was one of the first books that addressed this area [20]. It started a development that has had a large impact on both the industrial and the academic fields. He then published many more books on the subject, often teaming with P. Dewhurst.

The DfM tools are often very useful, but they normally focus on the later stages of the design process where the main costs already have been decided. Therefore, it is important to combine a concurrent and integrated design process with the use of design for manufacturing/assembly tools. The two approaches complement each other.

Two researchers that have had a great impact on product design and development are Ulrich and Eppinger [17]. They stress the importance of design for manufacturing. Their DfM method includes the following steps:

- Estimate the manufacturing costs
- Reduce component costs
- Reduce assembly costs

- Reduce the manufacturing support costs
- Consider the impact of DfM decisions on other factors

The biomechatronic methodology can be combined with this DfM method, for example, for estimation of consequences that different alternative concepts have for the costs. The methodology can act as a support for the task of performing these estimations.

7.5 CONCLUSIONS

The biomechatronic design methodology has, as discussed in this chapter, been applied on a device that exemplifies a composite of biology, biochemical and microbial conversion, pharmaceutical ingestion, radionuclide physics, micromechanics, medical diagnostics, and measurement technology. The breadth of techniques applied within the same device with its supporting consumables is apparent. A consequence of this is that the design alternatives and the possibilities to generate solutions are numerous. This strongly justifies mobilization of efficient design tools.

This example should have highlighted the utility of the mechatronic design methodology for a distinctly biology-driven device. The preference for one design solution is revealed early and further confirmed with the different design tools applied.

In this case, an already existing device has been selected to illuminate the advantages of the methodology. Not so surprising, the product development done in these UBT systems has today reached a stage of completion that this basic example confirms.

On the one hand, the generated analysis shows that the selection of components has been done wisely within the boundaries given. On the other hand, the results also suggest that the anatomy can still be refined, provided new components are invented.

REFERENCES

1. Qureschi, W.A., Graham, D.Y. (1997) Diagnosis and management of *Helicobacter pylori* infection. *Clin. Cornerstone* 1, 18–28.
2. Brown, L.M. (2000) *Helicobacter pylori,* epidemiology and routes of transmission. *Epidemiol. Rev.* 22, 283–297.
3. Peura, D.A. (1995) *Helicobacter pylori,* a diagnostic dilemma or a dilemma of diagnostics. *Gastroenterology* 109, 313–315.

4. Cirak, M.Y., Akyön, Y., Mégraud, F. (2007) Diagnosis of *Helicobacter pylori*. *Helicobacter* 12(Suppl. 1), 4–9.
5. Krogfelt, K.A., Lehours, P., Megraud, F. (2005). Diagnosis of *Helicobacter pylori* infection. *Helicobacter* 10(Suppl. 1), 5–13.
6. Rautelin, H., Lehours, P., Megraud, F. (2003) Diagnosis of *Helicobacter pylori* infection. *Helicobacter* 8(Suppl. 1), 13–20.
7. Zagari, R.M., Bazzoli, F. (2003) *Helicobacter pylori* testing in patients with peptic ulcer bleeding. *Digest. Liver Dis.* 35, 215–216.
8. Lee, J.M., Breslin, N.P., Fallon, C., O'Morain, C.A. (2000) Rapid urease tests lack sensitivity in *Helicobacter pylori* diagnosis when peptic ulcer disease presents with bleeding. *Am. J. Gastroenterol.* 95, 1166–1170.
9. Hegedus, O., Ryden, J., Rehnberg, A.S., Nilsson, S., Hellstrom, P.M. (2002) Validated accuracy of a novel urea breath test for rapid *Helicobacter pylori* detection and in-office analysis. *Eur. J. Gastroenterol. Hepatol.* 14, 513–520.
10. Borody, T.J., Wettstein, A.R., Campbell, J., Torres, M., Hills, L.A., Herdman, K. J., Pang, G, Ramrakha, S. (2008) Rapid and superior diagnosis of *H. pylori* infection by ^{14}C-urea Heliprobe™ test versus the PYtest®. *Gastroenterology*, 134(4), Supplement 1, A-329.
11. Gisbert, J.P., Pajares, J.M. (2004) ^{13}C-urea breath test in the diagnosis of *Helicobacter pylori* infection: a critical review. *Aliment Pharmacol. Ther.* 20, 1001–1017.
12. Yañez, P., Madrazo-de la Garza, A., Pérez-Pérez, G., Cabrera, L., Muñoz, O., Torres, J. (2000) Comparison of invasive and noninvasive methods for the diagnosis and evaluation of eradication of *Helicobacter pylori* infection in children. *Arch. Med. Res.* 31, 415–421.
13. Özdemir, E., Karabacak, N.I., Degertekin, B., Cirak, M., Dursun, A., Engin, D., Ünal, S., Ünlü, M. (2008) Could the simplified ^{14}C breath test be a new standard in noninvasive diagnosis of *Helicobacter pylori* infection? *Ann. Nucl. Med.* 22, 611–616.
14. Jonaitis, L.V., Kiudelis, G., Kupčinskas, L. (2007) Evaluation of a novel ^{14}C-urea breath test "Heliprobe" in diagnosis of *Helicobacter pylori* infection. *Medicina (Kaunas)* 43(1), 32–35.
15. Kuloglu, Z., Kansu, A., Kirsaclioglu, C.T., Üstündag, G., Aysev, D., Ensari, A., Kücük, N.Ö., Nurten, G. (2008) A rapid lateral flow stool antigen immunoassay and ^{14}C-urea breath test for diagnosis and eradication of *Helicobacter pylori* infection in children. *Diagn. Microbiol. Infect. Dis.* 62, 351–356.
16. Öztürk, E., Yesilova, Z., Ilgan, S., Arslan, N., Erdil, A., Celasun, B., Özgüven, M., Daglap, K., Ovali, Ö., Bayhan, H. (2003) A new practical low-dose 14C urea breath test for the diagnosis of *Helicobacter pylori* infection: clinical validation and comparison with standard tests. *Eur. J. Nucl. Med. Mol. Imag.* 30, 1457–1462.
17. Ulrich, K.T., Eppinger, S.D. (2008) *Product Design and Development*, 4th edition, McGraw-Hill, New York.

18. Derelöv, M., Detterfelt, J., Björkman, M., Mandenius, C.F. (2008) Engineering design methodology for bio-mechatronic products. *Biotechnol. Prog.* 24, 232–244.
19. Hubka, V., Eder, W.E. (1988) *Theory of Technical Systems, A Total Concept Theory for Engineering Design*, Springer-Verlag, Berlin.
20. Boothroyd, G., Poli, C., Murch, L.E. (1982) *Automatic Assembly*, Marcel Dekker Inc.

8

Microarray Devices

This chapter discusses the design of microarray devices. First, the principles and historical development of the microarrays are described and which methods predominates the applications (Section 8.1). Then, the needs of analysis are specified (Section 8.2). Finally, the design of microarrays for these needs is evaluated using the biomechatronic methodology (Sections 8.3).

8.1 PRINCIPLES, METHODS, AND APPLICATIONS OF MICROARRAYS

8.1.1 Principles and Technology

During the mid-1990s, oligonucleotide chips, or DNA chips in short, developed as a new methodology of analyzing genes from a biological sample on a small surface. The novelty of the invention was that for the first time it became possible to analyze many genes in parallel [1,2]. The very large number of cellular genes that the whole genome sequencing processed in the cell, opened up new experimental possibilities paving the way for functional genomic and transcriptomoic analysis. The variety of applications, especially in medicinal

Biomechatronic Design in Biotechnology: A Methodology for Development of Biotechnological Products, First Edition. Carl-Fredrik Mandenius and Mats Björkman.

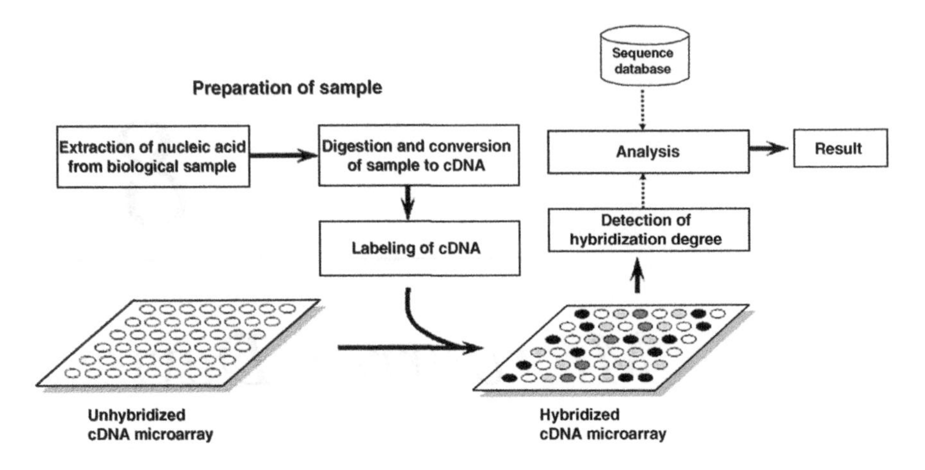

Figure 8.1 *The cDNA microarray analysis procedure: sample preparation, labeling, hybridization of microarray, and analysis of array data versus a sequence database. Normally, several arrays are analyzed consecutively and the relative responses are compared.*

research and diagnosis, became a driver of the development of DNA chips. The small size and multiplexing capacity of the chips coined the term DNA microarray preferred today.

The basic principle of the DNA microarrays is that oligonucleotides with specific and known sequences are immobilized as small spots in two dimensions on a surface in fixed positions (Figure 8.1). A sample consisting of a mixture of complementary DNA (cDNA) strands is brought in contact with the arrayed oligonucleotides on the chip surface. The DNA strands in the sample and DNA strands bound to the surface are allowed to hybridize. This occurs due to controlled hybridization conditions. The surface is washed removing unbound sample DNA strands. The hybridized strands remain on the spots where there is a complementary oligonucleotide. These spots can be identified. The sequences of the addressable spots are known. By that it is possible to tell which gene sequences are present in the sample.

Figure 8.2 summarizes the most commonly applied technologies of microarray methods [3].

8.1.2 Fabrication Methods

Fabrication of microarrays is based either on the synthesis of the oligonucleotides directly on the chip or on the delivery of presynthesized oligos onto the chip. When the synthesis method is applied, each unique nucleotide sequence on the microarrays is built up by stepwise chemical reactions on the surface, that is, the so-called *in situ* synthesis. This could be done with one nucleotide per step or by using blocks of nucleic acids and other biopolymers.

With each round of synthesis, nucleotides are added to growing oligonucleotide (oligo) of a desired length.

In another fabrication method based on delivery of presynthesized oligos, the exogenous deposition techniques is used onto the chip. The cDNA oligos are amplified by a polymerase chain reaction (PCR) and purified, and small quantities of DNA oligos are deposited onto known and addressable positions.

The key technical attributes for comparing the fabrication include microarray density and design, biochemical composition and versatility, reproducibility, throughput, quality, cost, and ease of prototyping. During the 1990s, mainly three fabrication methods emerged, which were used in automated microarray production (Figure 8.2).

The commercially most successful method has been the *photolithographic method* developed by Affymetrix (see also below) [4]. The method combines photolithography, adopted from the semiconductor industry, with DNA synthesis chemistry, for preparing high-density oligonucleotide microarrays (Figure 8.2a). A key advantage of this approach over nonsynthetic methods is that photoprotected derivatives of the four nucleotide bases allow chips to be manufactured directly from sequence databases, thereby removing the

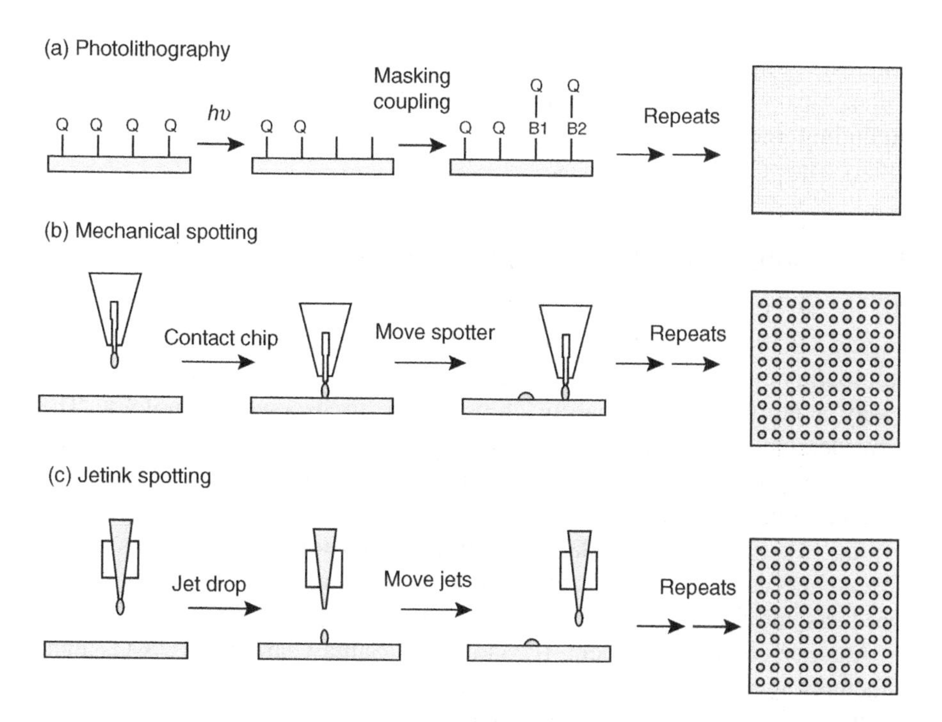

Figure 8.2 *cDNA three methods for fabrication of DNA microarrays chips. (a) Photolithography chemistry. (b) Mechanical spotting of preprepared cDNA probes. (c) Jet ink spotting of the chip surface. Adapted and reproduced with permission from [3].*

uncertain and burdensome aspects of sample handling and tracking. Another advantage of the photolithographic approach is that the use of synthetic reagents minimizes chip-to-chip variation by ensuring a high degree of precision in each coupling cycle. One disadvantage of this approach is, however, the need for photomasks, which are expensive and time consuming to design and build.

Another approach for fabrication is mechanical microspotting (Figure 8.2b). Microspotting, a miniaturized version of earlier DNA spotting techniques, encompasses deposition of premade biochemical substances on solid surfaces by printing small quantities, to enable automated microarray production. Printing is accomplished by direct surface contact between the printing substrate and a delivery mechanism that contains an array of tweezers, pins, or capillaries that serve to transfer the biochemical samples to the surface.

A third fabrication method for microarrays is the "drop-on-demand" delivery approach (Figure 8.2c). The most elaborate version of the method is ink jetting, a technique that utilizes piezoelectric and other forms of propulsion, to transfer biochemical substances from miniature nozzles to a solid surface. Similar to the microspotting approach, drop-on-demand technologies allow high-density gridding of virtually any biomolecule of interest, including cDNAs, genomic DNAs, antibodies, and small molecules. Ink jetting-based methods are developed at several centers, including Incyte Pharmaceuticals (Palo Alto, CA, USA) and Protogene (Palo Alto, CA, USA).

8.1.3 Companies Developing Microarrays

Today, several manufacturers of DNA microarrays compete on the market. Figure 8.3 shows few of these. One of the pioneering companies developing microarrays was Affymetrix (Santa Clara, CA, USA). The company offers today a number of microarray chips based on its own technology. The considerable flexibility of its products and manufacturing methods allows the design of the arrays to be tailored for their intended use, such as whole-genome transcriptome mapping, gene expression profiling, or custom genotyping [5]. Except its standard assortment, a custom service exists where researchers can design their own arrays for organisms not covered by existing products and for specialized or directed studies [5].

Another company, Agilent Technologies (Santa Clara, CA, USA), manufactures a variety of customized oligonucleotide microarrays for use in their multiple two-color microarray applications [6]. Agilent manufactures microarrays via an *in situ* synthesis procedure based on inkjet printing of nucleotide precursor molecules and chemical processing of each added nucleotide layer.

The typical probe length is 60 nucleotides. The method is effectively 5-ink (4 bases plus catalyst), 60-layer printing with reregistration at each layer. Because the technology synthesizes oligonucleotides on demand, based on a digital representation of the desired sequences and layout, it is well suited to manufacturing custom microarray designs or microarrays that mix standard and custom probe content. Typically, the maximum feature count is 184,672 per 1×3 in. slide [6].

Other major companies that develop and supply DNA microarrays are Eppendorf, GE Healthcare, and Biodot (see also examples in Figure 8.3).

8.1.4 Applications of DNA Microarrays

Microarrays are now being used for a number of applications, for example, genome-wide expression monitoring, large-scale polymorphism screening, and mapping, and evaluation of drug candidates.

Gene expression information is, for example, used to support characterization and validation of drug targets, to select viable drug candidates, and to predict potential toxicity effects [7,8].

In particular, the use of microarray gene expression information can support lead selection by using gene expression profiles for test compounds against known toxins. The effects can also be compared with the effects of existing marketed pharmaceutical products, to support the selection of compounds with similar mode of action [9].

High-density microarrays will allow the measurement of the gene expression component of disease and the identification of promising new drug targets.

Recent applications supported by microarrays are used to acquire critical information for the drug discovery process, such as mapping disease genes in subgroups of the patient populations for allele-specific therapeutics. Also, global genetic information will be essential for more effective treatment [7].

With the availability of complete genome sequences of an increasing number of different species, antimicrobial drug discovery has the opportunity to access a remarkable diversity of genomic information [10].

In addition, using DNA microarrays makes it possible to utilize primary sequence data for microbial pathogens for measuring levels of transcription and for detecting polymorphisms. By measuring the bacterial host mRNA expression, it is possible to assess the functions of uncharacterized genes, physiological adaptation, virulence-related genes, and effects of drug treatment [11].

Microarray analysis provides a way to connect genomic sequence information and functional analysis. Recent experiments involving the use of cDNA

Microarray device	Analysis System	Vendor
Affymetrix gene chips 	The gene chip is used With an analyzer station for chip processing, hybridization and analysis, including advanced software for analysing data.	Affymetrix (USA)
Agilent DNA Microarray 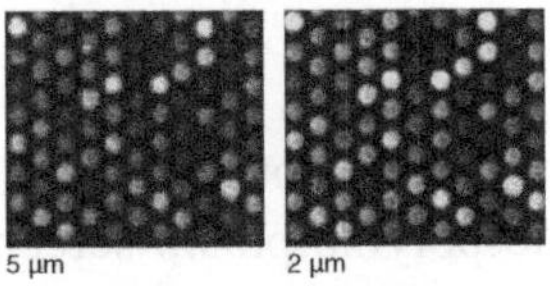	The Agilent DNA microarray is scanned in system that can read glass slide microarrays, and analyze them using an extraction software.	Agilent Technologies (USA)
GE Healthcare microarray imager 	The microarray imager (Typhoon) uses phosphor autoradiography technology with four-color, non-radioactive fluorescent labeling techniques. The scanning of the microarray is controlled and analyzed by a software.	GE Healthcare (S)
Biodot dispense system 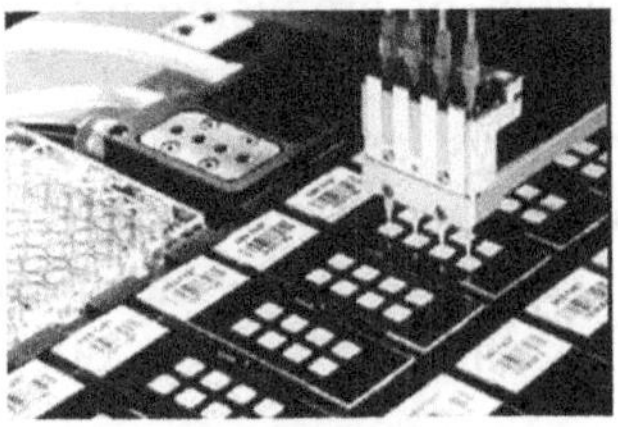	The dispense system is based on a precision XYZ platform with 4-20 nL air jet dispense channels with micropumps, ultrasonic washing and PC control.	Biodot Inc, (USA)

Figure 8.3 *Four commercial microarray products (Affymetrix, GE Healthcare, Agilent, Eppendorf, and Biodot). Adapted and reproduced with permissions from Affymetrix, Agilent, GE Healthcare, and Biodot.*

microarrays for expression monitoring in strawberry and petunia indicate the immediate applicability of DNA chips in agricultural biotechnology.

Microarrays can also be used for functional analysis, to help better understand the fundamental mechanisms of plant growth and development with links to effects of hormone and herbicide treatment, and genetic background and environmental condition.

Interaction of genes and environment is also of particular importance in plants, and is another area of interest. The microarrays can provide to plant biotechnology companies a rapid analysis of transgenic plants. These data can allow to establish correlations between expression and a host of desirable traits such as fertility, yield, and resistance to environment and insects. Microarrays may help both to understand effect of small molecules and by that accelerate discovery of herbicides and to analyze mechanism of action [12].

Other examples of application are found in oncology gene research, for example, for detecting abundance in point mutations for diagnosis of colorectal cancers [13].

Ways to increase capacity of array platforms for specific applications are constantly sought, such as assays of gene expression, proteomics, genotyping, DNA sequencing, and fragment analysis [14].

8.2 SPECIFICATION OF NEEDS

The needs and target specifications for DNA microarrays are amply discussed in the bioengineering literature.

Typical attributes of interest to elaborate on and control in the design concern many factors [8,15–18]. Basically, these attributes include the target preparation, spotting procedure, sample preparation, hybridization conditions, scanning of the array, and data interpretation.

Concerning the target preparation, typical issues are suitable length of the nucleotide target, target type (e.g., single-stranded DNA, double-stranded DNA, oligos, RNA, and PNA), costs, preparation time, target selection, and controls.

The spotting procedure involves linkage to the matrix, surface materials, density of spots, reproducibility, target concentration, cost of spotting preparation, array preparation time, and automation degree.

The preparation of the sample to be analyzed should be fluorescent labeled, could use multiply dyes, could involve direct labeling, could concern mRNA or total RNA, must have reproducibility of a sufficient level, and must be possible to quality control.

Hybridization should cope with a linear range, should have proper hybridization conditions, show a good signal-to-noise ratio, and should have high specificity.

Scanning of the array should probably be compatible with the use of multiple dyes and have a certain resolution, sensitivity, speed, and degree of automation.

TABLE 8.1 Specification of Needs and Target Specifications for a DNA Microarray

Needs → Metrics	Target Values	Units
Target preparation		
Correlate with real-time SPR mRNA	Not lower than 0.96	Correlation factor
Correlate with existing microarrays	Not lower than 0.90	Correlation factor
Correlation with protein expression	Not lower than 0.50	Correlation factor
Lot variation of consumables low	Not higher than 2%	% variation
Comply with recommendations from standardization organizations	Comply with at least WHO guidelines	Intern. organizations
Possible to calibrate with standards	Yes	Yes/no
No false negative	Yes	Yes/no
Low false positive	<3%	Incidence
Precision high	>95%	% precision
Accuracy high	>96%	% accuracy
The microarray		
Limited interference	<5%	% interference
Short sampling preparation time	<24 h	Hours
Convenient operation temperature	15–35°C	°C
Instrumentation station		
Small sample volume required	50	μL
Convenient instrument size	L 80–120 mm/H 8–20 mm	L/H aspects
Low price per microarray	<0.5 EUR	Top acceptable price
Low-cost reagents	<10% of array cost	
Short run time per array	2 h	Occupation of instrument
Evaluation software		
User-friendly	High	High/medium/low
Support provided in short time	<1 day	Days
Consumables delivery time short	<1 day	Days

Finally, interpretation of the generated data must use software programs that can link to a sequence database and the software should have a suitable availability, a good graphic interface, allow multiple comparisons, support statistics, and fit into a laboratory organization.

Table 8.1 compiles these attributes and suggests target values for the design. As before, these values are set up by a design team as goals. The rationale for the values might be from looking at competitors' microarrays or from a thorough analysis of the customers' actual needs.

8.3 DESIGN OF MICROARRAYS

8.3.1 Generation of cDNA Microarray Concepts

Section 8.1 gave ideas about alternative DNA microarray concepts. These concepts, realized in commercial products, were designed driven by

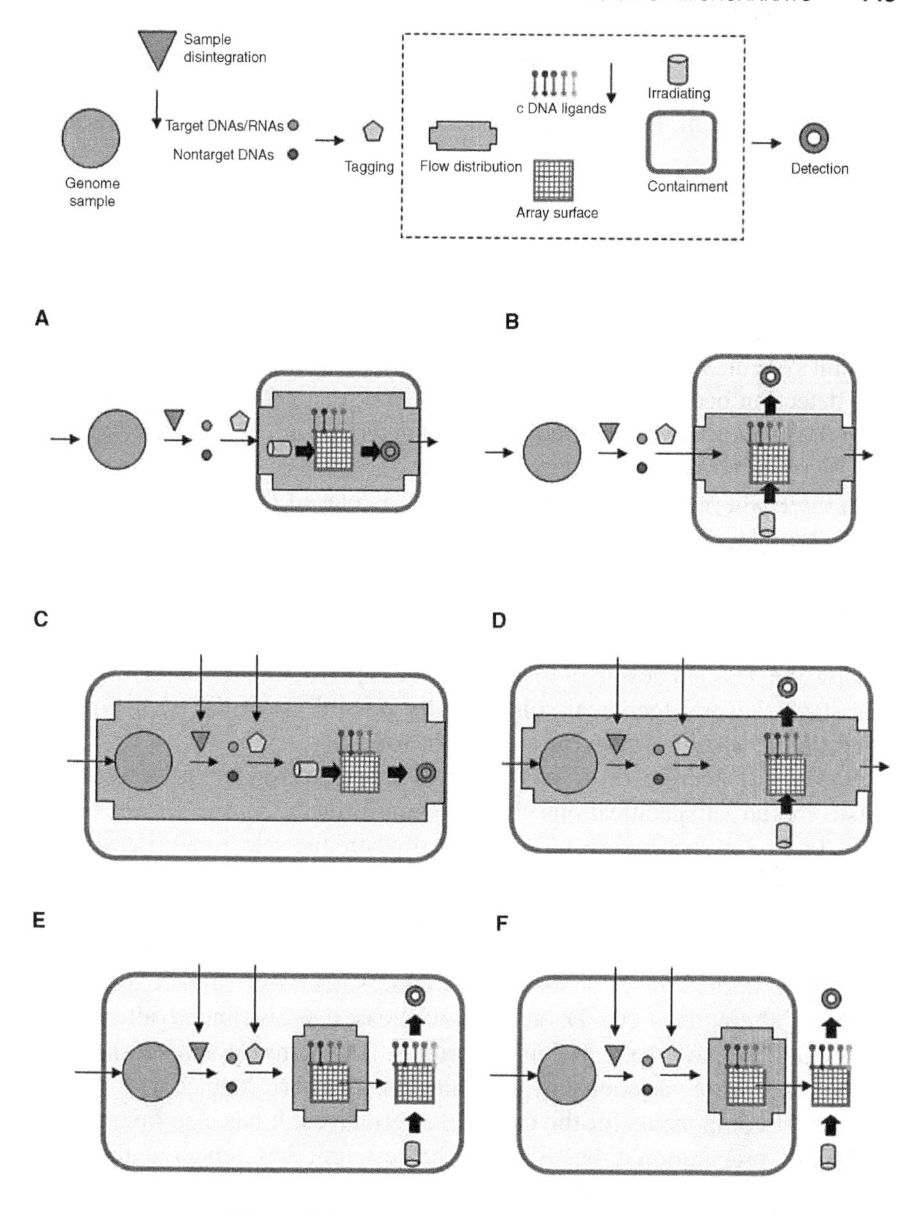

Figure 8.4 Concepts generated for microarrays.

technology possibilities. This is common for completely new inventions as the DNA microarrays show.

In the concept generation chart of Figure 8.4, we have tried to take a slightly different angle.

The alternative A shows a concept where the sample is disintegrated into short oligos and tagged outside the container with the microarray. The preparation is then purged into a container with the array, a flow distributor, irradiator, and detector. The hybridization takes place inside the flow distributor. Used sample is perfused out from the container at a set interval.

Alternative B is equivalent to A except that the irradiation and recording of the microarray happens outside the flow distributor. This concept may represent system where the microfluidics is separated from sensor electronics.

Alternative C is a system where all steps are carried out inside the flow conduit system. Sample and reagents are added to the flow distribution system. The detection occurs inside as well.

In the D alternative, the irradiation and detection are again placed outside the microfluidic system.

In the E alternative, only the microarray is placed in the flow distributor while sample preparation is carried out in the contained part while the detection of the array is done by separating it from the flow distributor and record its radiation inside the container.

In F, this recording takes place outside the container.

How the detailed design of the alternative is performed is not prescribed at this stage. Several technical solutions are possible with different types of microfluidic systems, detectors, and light sources.

Before considering these possibilities, an evaluation of the alternatives versus the target specifications should be performed.

In Table 8.2, such a screening is shown where the selection criteria cover most of the essential target specifications.

For example, for the target specification criteria, the correlation aspects have in most values been considered to require prototyping for evaluation. For one criterion, high precision, the assessment has favored the more compact alternatives (C, D, and E) whereas the distributed alternatives (A, B, and F) have been assigned a minus. The same considerations have been made for lot variations, by assuming a better performance with a lesser number of components for the compact alternatives. It has also favored size and short preparation time. Size of the test unit has rendered the same assessment. For several other criteria, it has been difficult to assess without experimental testing performance or other properties. Thus, it has been assigned '0' values.

This rough screening suggests that the alternatives D and F are first choices, while F is closely second, and the remaining of lesser ranking.

As stated in previous chapters, the concept screening is only indicative and should preferably be followed up by a more thorough scoring matrix evaluation. We omit this step in this chapter and move directly to the Hubka–Eder mapping with a focus on the more compact alternatives.

TABLE 8.2 Concept Screening Matrix for Selection Criteria

	Concepts					
Selection Criteria	A	B	C	D	E	F
Target preparation						
Correlate with real-time SPR mRNA	+	+	0	0	0	0
Correlate with existing microarrays	0	0	0	0	0	0
Correlation with protein expression	–	–	–	–	–	–
Lot variation of chip low	0	0	+	+	+	+
Comply with recommendations from standardization organizations	0	0	0	0	0	0
Possible to calibrate with standards	+	+	+	+	+	+
No false negative	0	0	0	0	0	0
No false positive	0	0	0	0	0	0
Precision high	–	–	+	+	+	–
Accuracy high	0	0	0	0	0	0
The microarray	0	0	0	0	0	+
Limited interference	+	+	+	+	+	+
Short sampling preparation time	0	0	+	+	+	+
Convenient operation temperature	+	+	+	+	+	+
Instrumentation station	–	+	0	+	+	+
Small sample volume required	+	+	+	+	+	+
Convenient instrument size	0	0	+	+	+	+
Low price per microarray	+	+	+	+	+	+
Low-cost reagents	0	0	+	+	+	+
Short run time per array	0	0	0	0	0	0
Evaluation software	0	0	0	0	0	0
User-friendly	+	+	+	+	+	+
Support provided in short time	+	+	+	+	+	+
Consumables delivery time short	+	+	+	+	+	+
Sum +'s	9	10	10	14	14	14
Sum 0's	12	12	13	9	9	8
Sum –'s	3	2	1	1	1	2
Net score	6	8	9	13	13	12
Rank	**6**	**5**	**4**	**1**	**1**	**3**

8.4 DESCRIPTION OF THE SYSTEMS INVOLVED IN THE DESIGN CONCEPTS

The transformation process (TrP) of the microarray analysis is shown in the Hubka–Eder map in Figure 8.5 together with systems involved in the design concept. The sample is the key ingoing operand and secondary inputs are the disposable microarray, reagents for preparation, and hybridization as well as sequence data. In the preparation phase, the oligos are prepared and the microarray is spotted or built with photolithography. In the executing phase, oligos in the sample are labeled and hybridized to the array, washed, and

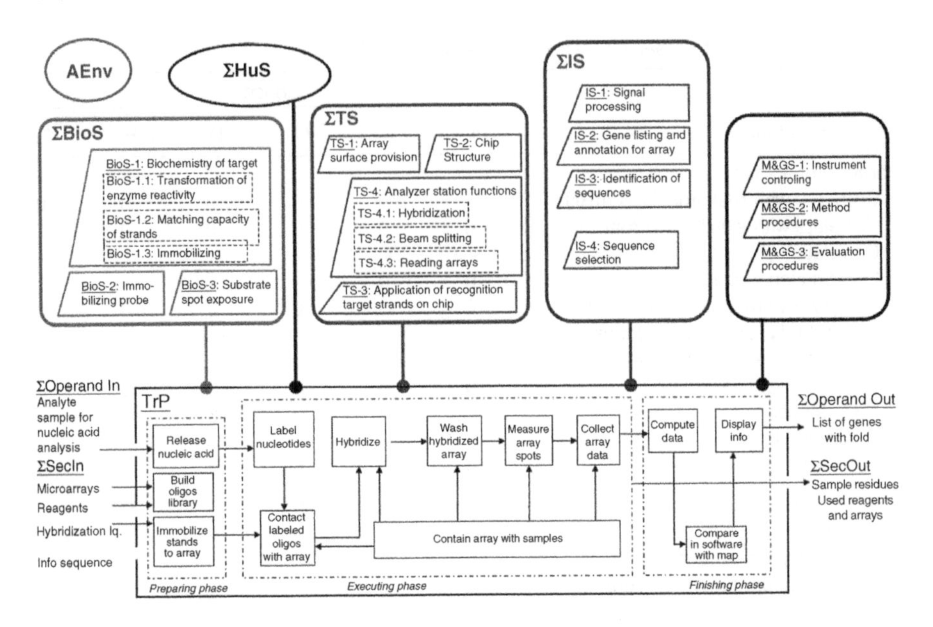

Figure 8.5 *Hubka–Eder map of a microarray with analysis station.*

measured. Also, the acquisition of binding data is included in the executing phase. In the finishing phase, recorded data are analyzed, sequence data are compared, and various algorithms for treating the data are applied. Results are displayed to the user in a form that is understandable from user perspective. Thus, the main outgoing operand is a list of data on mRNA content in the sample. Secondary outputs are used arrays and reagents.

The various systems needed to effectuate the transformation are discussed below.

8.4.1 Biological Systems

The biological and biochemical systems (ΣBioS) for carrying out the TrP are the ssDNA strands attached to the array with their functional capacity to recognize and hybridize to complementary single strands labeled in the sample. Thus, the conditions of the hybridization reaction and their realization on the microarray surface are key functions of the biological systems. Part of this is the spatial availability of the strands, their length, their mismatch, and their background and control responses. Most of these relations and functionalities are well known facts from the biochemistry of hybridization reaction. Critical factors to evaluate are the effects and interactions of the surface, spacers from grafting, microdiffusion in spots, and so on.

8.4.2 Technical Systems

The technical systems (ΣTS) to be used on the TrP are a combination of surface functions, irradiation and detection devices, and reaction chemistry to couple the inorganic part of the surface to the biological molecules, that is, the oligonucleotides. It is also the microfluidics, microelectronics integrated into the devices, and containment functions. Within the microfluidic functions, transportation of liquids is a key element. The detection function is possible to further consider here, using photochemistry, fluorescence, or electronic transduction of signals. The fabrication of the microarray surface as discussed in Figure 8.2 is part of the functionality provided by the technical system and subsystems. Thus, what appears as the dominating focus in the microarray literature is consequently just one of the important issues in the design of the microarray concept.

8.4.3 Information System

The information systems (ΣIS) involved are the signal processing, addressing the spot (coordinates) on the array, the recorded chip, and the process of analyzing the data versus sequence database information. This is done with the help of software algorithms and procedures that the software prescribes based on bioinformatics theory and application. This part is today a science of its own, partly owing to the access to the array technology and also due to intrinsic knowledge about biostatistical methods.

8.4.4 Management and Goal Systems and the Human Systems

The management of the whole microarray analysis is probably embedded in the expert system, running the steps in the microarray transformation process, initiating analysis, and handling data inputs from the operator/scientist running the analysis.

The role of the humans is significant here. It is the performer of the analysis that must be able to manage the system. Thus, the ΣHuS entity is more prominent here. It is probably a team of scientists or clinicians that are going to use the microarray for a particular analytical purpose/goal.

The ΣM&GS and the ΣHuS thus become very interactive in the performance of the analysis. It is rather obvious that the utility of the microarray strongly depends on its operability in this context.

8.4.5 Active Environment

The active environment (AEnv) includes the biological sample. Here, the variation in the cell and the cellular responses becomes an integral part of the unforeseeable AEnv. It is rather obvious that the fate of oligonucleotides from the cell may, as per our present understanding of these molecules, be under the influence of unknown factors. This emphasizes the need of proper control (blanking) procedures to handle this AEnv.

8.4.6 Interaction Analysis

The function interaction matrix analysis is here as motivated as in the previous cases. Figure 8.6 shows a complete Functions Interaction Matrix that includes all the systems and the TrP with subsystems as presented in the Hubka–Eder map of Figure 8.5.

The FIM rating confirms several of the statements made above. The interactions between the microarray surface and the biochemical reactions during hybridization are important to consider in the design. Thus, a design alternative that carefully controls these steps would be favorable for reaching the targets in the specification concerning accuracy and specificity.

Main Systems		ΣBioS					ΣTS				ΣIS			ΣM&GS			ΣHuS		
	Subsystems	Enzyme reactivity	Match capacity	Immobilizing array	Coupling sample strand	Substrate spot exposure	Array surface provision	Chip sgtructure	Aanalyzer station	Substrate of sample strand	Signal processing	Gene listing and annotation	Identification of sequence	Instrument control	Methods procedural	Evaluation procedures	Operation of microarray	Compuation and analysis	Support of customer
Σ BioS	Enzyme reactivity		3	4	4	1	1	2	2	3	2	2	2	2	5	5	3	3	1
	Match capacity	1		2	2	1	4	4	4	4	4	2	2	4	4	5	4	5	1
	Immobilizing array	3	4		4	5	5	5	5	5	4	4	2	4	4	4	4	2	1
	Coupling sample strand	4	4	4		4	4	4	4	4	2	2	2	4	4	4	2	2	1
	Substrate spot exposure	1	5	5	5		5	5	5	5	5	5	5	5	4	5	4	4	1
Σ TS	Array surface provision	1	5	5	5	5		5	5	5	4	4	4	4	5	5	1	1	1
	Chip structure	1	4	5	5	4	4		4	4	4	1	1	4	4	4	1	1	1
	Analyzer station function	1	4	5	4	5	5	5		5	5	5	5	5	5	5	1	1	1
	Substrate of probe strand	1	4	5	5	5	5	5	5		5	5	5	5	3	5	1	1	1
Σ IS	Signal processing	1	4	5	3	1	1	1	1	1		5	5	5	3	5	1	1	1
	Gene listing and annotaion	1	4	3	3	1	1	1	1	1	1		5	1	1	5	1	1	1
	Identification of sequence	1	4	3	3	1	1	1	1	1	1	1		1	3	5	1	1	1
Σ	Instrument controlling	5	5	5	5	5	5	5	5	5	5	5	5		5	5	5	5	1
	Methods procedural	4	4	4	4	4	4	4	4	4	4	4	4	4		4	4	4	5
	Evaluation procedures	1	1	1	1	1	1	1	1	1	1	5	5	5	5		4	5	5
Σ HuS	Operation of microarray	3	3	3	3	3	4	4	4	4	1	2	2	5	5	4		5	5
	Computation and analysis	1	1	1	1	1	3	3	3	3	1	5	5	5	5	5	5		5
	Support of micro-array customer	1	1	1	1	1	1	1	1	1	1	1	1	5	5	5	5	5	

Figure 8.6 Functions Interaction Matrix for Microarray.

The importance of the fabrication technology of the microarray is visible. The interactions that are involved in this are the properties of the materials, devices of spotting, and the liquid properties of the solutions to be transformed or transported. In the choice of techniques, detailed FIM ranking can reveal preferences of available techniques for spotting jet ink application of photolithographic alternatives.

8.5 CONCLUSIONS

Much of the cDNA chip development has been done from a design engineering perspective where, as it appears in the literature, the design approach has been systematic and structural. The examples given in this chapter have been collected from presentations and reports from several of the SMEs that are behind the efficient development of new microarray devices. We believe that the involvement of engineering from the fields of micromechanics and microelectronics has been pivotal for to this development. If so, this chapter should be regarded as a retrospect confirmation of a sound design approach for a biotechnology product. It follows truly the biomechatronic concept by uniting functional biological molecules (nucleotides) and mechatronics (the chips).

This chapter has entirely dealt with DNA microarray design issues. However, microarray technology encompasses additional opportunities for parallelized biointeraction-based analysis. Of particular relevance are protein microarrays with the expressed proteins that are analyzed by using immunorecognition or other molecular recognition mechanisms [19–22]. The combination of mRNA data, from a DNA microarray, with the data for the processed mRNA product brings the full picture of the expression profile. The power to analyze all expressed proteins on the same chip is undeniable.

Other ascents in the application of chips are steps toward whole cells, for example, cellular chips [23].

The market success of microarrays should stimulate the development of new design variants and further development of existing methods. The biomechatronic design approach is in that context of particular help.

REFERENCES

1. Hoheisel, J.D. (1997) Oligomer-chip technology. *Trend. Biotechnol.* 15, 465–469.
2. Ramsay, G. (1998) DNA chips: state of the art. *Nat. Biotechnol.* 16, 40–44.

3. Schena, M., Heller, R.A., Theriault, T.P., Konrad, K., Lachenmeier, E., Davis, R.W. (1998) Microarrays: biotechnology's discovery platform for functional genomics. *Trend. Biotechnol.* 16, 301–306.
4. Fodor SPA, Rend JL, Pirrung MC, Stryer L, Tsai Lu A, Solas D (1991) Light-directed, spatially addressable parallel chemical synthesis. *Science* 251, 767–773.
5. Dalma-Weiszhausz, D.D., Warrington, J., Tanimoto, E.Y., Miyada, C.G. (2006) The Affymetrix GeneChip™ platform: an overview. *Meth. Enzymol.* 410, 3–28.
6. Wolber, P.K., Collins, P.J., Lucas, A.B., De Witte, A., Shannon, K.W. (2006) The Agilent *in situ*-synthesized microarray platform. *Meth. Enzymol.* 410, 28–57.
7. Braxton, S., Bedilion, T. (1998) The integration of microarray information in the drug development process. *Curr. Opin. Biotechnol.* 9, 643–649.
8. Lennon, G.G. (2000) High-throughput gene expression analysis for drug discovery. *Drug Discov. Today* 5, 59–66.
9. Loferer, H. (2000) Mining bacterial genomes for antimicrobial targets. *Mol. Med. Today* 6, 470–474.
10. Cummings, C.A., Relman, D.A. (2000) Using DNA microarrays to study host–microbe interactions. *Emerg. Infect. Dis.* 6, 513–525.
11. Lemieux, B., Aharoni, A., Schena, M. (1998) Overview of DNA chip technology. *Mol. Breeding* 4, 277–289.
12. Soper, S.A., Hashimoto, M., Situma, C., Murphy, M.C., McCarley, R.L., Cheng, Y. W., Barany, F. (2005) Fabrication of DNA microarrays onto polymer substrates using UV modification protocols with integration into microfluidic platforms for the sensing of low-abundant DNA point mutations. *Methods* 37, 103–113.
13. Ricco, A.J., Boone, T.D., Fan, Z.H., Gibbons, I., Matray, T., Singh, S., Tan, H., Tian, T., Williams, S.J. (2002) Application of disposable plastic microfluidic device arrays with customized chemistries to multiplexed biochemical assays. *Biochem. Soc. Trans.* 30, 73–78.
14. van Hal, N.L.W., Vorst, O., van Houwelingen, A.M.M.L., Kok, E.J., Peijnenburg, A., Aharoni, A., van Tunen, A.J., Keijer, J. (2000) The application of DNA microarrays in gene expression analysis, *J. Biotechnol.* 78, 271–280.
15. Yoshino, M., Matsumura, T., Umehara, N., Akagami, Y., Aravindan, S., Ohno, T. (2006) Engineering surface and development of a new DNA microarray chip. *Wear* 260, 274–286.
16. Hughes, T.R., Shoemaker, D.D. (2001) DNA microarrays for expression profiling. *Curr. Opin. Chem. Biol.* 5, 21–25.
17. Waliraff, G. M., Huisberg, W. D., Brock, P. J. (2000) Lithographic techniques for the fabrication of oligonucleotide arrays. *J. Photopolym. Sci. Technol.* 13, 551–558.
18. Situma, C., Hashimoto, M., Soper, S.A. (2006) Merging microfluidics with microarray-based bioassays. *Biomol. Eng.* 23, 213–231.

19. Mitchell, P. (2002) A perspective on protein microarrays. *Nat. Biotechnol.* 20, 225–229.
20. Stears, R.L., Mantinsky, T., Schena, M. (2003) Trends in microarray analysis. *Nat. Med.* 9, 140–145.
21. Weller, M.G. (2003) Classification of protein microarrays and related techniques. *Anal. Bioanal. Chem.* 375, 15–17.
22. Oh, S.J., Hong, B.J., Choi, K.Y., Park, J.W. (2006) Surface modification for DNA and protein microarrays. *Omics* 10, 327–343.
23. Kolchinsky, A.M., Gryadunov, D.A., Lysov, Y.P., Mikhailovich, V.M., Nasedkina, T.V., Turygin, A.Y., Rubina, A.Y., Barsky, V.E., Zasedatelev, A.S. (2004) Gel-based microchips, history and prospects. *Mol. Biol. (Russ.)* 38, 4–13.

9

Microbial and Cellular Bioreactors

The chapter begins with an update of the development of bioreactors during the 1970s–1990s (Section 9.1). The typical needs of the bioreactor users, whether they are academic researchers or they are industrial manufacturers of bioproducts, are discussed and specified (Section 9.2). The conventional bioreactor models that these needs have resulted in are analyzed from a biomechatronic design perspective. These are exemplified for two well-known bioreactor applications: fungal penicillin fermentation and protein production by mammalian cells (Section 9.3). Finally, recent trends in bioreactor design are analyzed and discussed with examples from micro-bioreactors and bioreactors for culturing tissue cells, stem cells, and plant cells (Section 9.4). Throughout the chapter, the Hubka–Eder mapping is applied for identifying the key issues of this versatile design topic.

9.1 BIOREACTOR DEVELOPMENT DURING THE 1970s–1990s

In a broad sense, bioreactors can be considered as any technical device that carries out biological reactions in a sufficiently controlled manner. Biological reactions can take place in a wide range of biomolecular and cellular systems: as single enzymes, bacterial and other microbial cells, animal cells,

Biomechatronic Design in Biotechnology: A Methodology for Development of Biotechnological Products, First Edition. Carl-Fredrik Mandenius and Mats Björkman.

mammalian cells, human cells and tissues [1,2], stem cells [3], gene vectors, or plant cells [4].

Preferably, the cell culture is a well-defined monoculture. However, in industrial practice, often multicultures are the actual biological systems due to impure culture handling, large-scale culture systems, or spontaneous culture changes. Sometimes multicellular systems are used intentionally, for example, in solid-state fermentations [5] and wastewater treatment processes.

Numerous bioreactor setups and configurations of various technical designs have been developed during the past decades. Industry has quite often propelled the development of bioreactor apparatuses, mostly with the purpose of using them in a particular manufacturing process.

This has resulted in a variety of bioreactor designs for diverse types of cells, such as airlift reactors, membrane reactors, hollow-fiber reactors, and bioreactors for microcarrier cultures and immobilized cells (Figure 9.1). The bioreactor developments during the 1970s–1990s are comprehensively described and reviewed in a number of books and articles (see, for example, Refs [6–8]).

In *food biotechnology* applications in particular, bioreactors exhibit a variety of configurations. The food industry has often remained with traditional and old-fashion bioreactor types. This is the case especially both in small-scale wine and beer production and in solid-state fermentation of food products [5]. Today, the food industry mostly uses modern bioreactors for large-scale production, in particular for dairy, wine, and other related food products. The safety requirements and cost margins in the food industry are different from, for example, drug manufacture. Thus, this requires critical attention in the design work.

In the *biopharmaceutical industry*, bioreactors must meet very high standards to comply with the requirements of the pharmaceutical regulatory authorities. The production facility must adhere to Good Manufacturing Practice (GMP) and allow proper validation procedures to be carried out. Normally, the purity of the drug products must be very high, something which cannot be reached in the bioreactor, so the bioconversion must adapt to the subsequent downstream process. The use of mammalian cell cultures and complex media has added new dimensions to the development. Together, this results in significantly more expensive bioreactor designs, which, as a consequence, emphasizes the need of finding economically optimal design alternative.

The *regulatory requirements* of the bioreactor design for other types of products, such as diagnostics, specialty biochemicals, industrial enzymes, and commodity chemicals, are less demanding. The bioreactor types for these bioproducts belong to the categories batch or continuous stirred, airlift, and tower bioreactors, all well-established and well-characterized apparatuses.

Several *commercial bioreactor products* have high visibility on the biotech product market. Companies such as Bioengineering, Applikon Biotechnology, Infors, Brunswick, Stedim-Sartorius (B Braun), and GE Healthcare (formerly Pharmacia Biotech) are well-known equipment developers and vendors, especially for applications in pharmaceuticals and diagnostics.

A rarer bioreactor application is the processing of *plant cell cultures* where callus tissues from trees, fruits, or herbs are adapted to grow in suspension or aggregated cultures for production of unique metabolites in the plant cells [4]. Recently, recombinant protein production in plant cells is also a new approach [17]. Plant cultures increase the demands on the bioreactor design since they grow to high densities and create viscous liquids of media and suspended cells. Diffusion of nutrients and gases in the media in the reactors and separation of formed products become critical for their operation and require that better design solutions are accomplished. Also, the long time span of the plant cell batches in the manufacture makes the sterility barrier a critical design issue.

In *industrial process R&D work*, especially in the biopharmaceutical companies, bioreactors are well established. To test and optimize the bioreactor operation, with respect to choice of strain or cell culture and to media composition, and to optimize critical process, variables before scale-up are economically necessary. Bioreactors for process development have become a special category of bioreactors, which need to be better equipped and suited for testing and optimization.

Research in *biochemical engineering science* has contributed significantly to the understanding and improvement of the fundamentals of bioreactor design. The research has uncovered important aspects of fluid dynamics of the bioreactor by using sophisticated sensors and advanced mathematical modeling methods [18,19,20–23]. Today, training at engineering schools and universities prepares new engineers for exploiting this understanding in design work in biotechnology industry [24]. This has paved the way for new conceptual bioreactor designs. These efforts have also been closely connected to progress in tissue engineering, stem cell research, and other fields.

In recent years, the *miniaturized bioreactors* have become popular tools for process development [25–27]. This scale-down approach has resulted in micro- or even nanoliter reactor volumes [14]. The miniaturized bioreactors usually serve as valuable tools for prestudies for the prediction of the conditions on a larger production scale.

Devices that mimic human organs, such as liver and kidney, are often referred to as bioreactors, or more precisely, *organ-simulating bioreactor devices*. The design of these devices is treated in Chapter 12 owing to their quite different aims, prerequisites, and demands. Mostly, these devices consist of multicellular systems, which make them more difficult to design and operate.

Another design concept that has drawn much attention is the *wave bioreactor* for single-use cultures [15]. The cells are confined in a plastic bag that is discarded after the batch process. Each disposable bag is equipped with necessary ports, connections, and sensors. It is delivered to the user presterilized. It is positioned

Bioreactor type	Characteristics	Dominating application	Ref
Stirred tank bioreactor 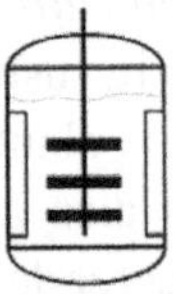	Homogeneous mixing Containment and in situ sterilizable Most common type in industry	All microbial and cell culture bioreactors up to 25-50 m^3.	[9]
Membrane bioreactor 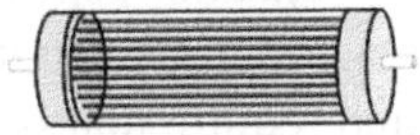	Hollow fiber bundles Small scale applications Difficult to reuse	Tissue cultures or surface growing cell cultures	[10]
Airlift bioreactor 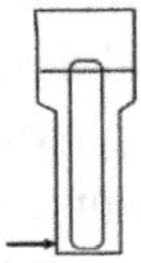	Suitable for microcarrier or flocculating cultures Expensive reactor design	Hybridoma cell cultivation with micro-carriers	[10, 11]
Bubble-column bioreactor 	Low mixing degree Low energy consumption	Large scale yeast manufacture	[12]
Draft tube bioreactors 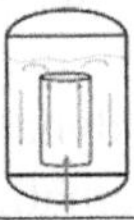	Expensive design solution	Methanol and single cell production	[12]
Mixer-settler bioreactor 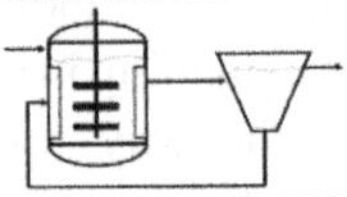	Expensive set-up Can be realized on very large scale 70 000 m^2 surface area	Wastewater treatment	[9]
Koji bioreactor 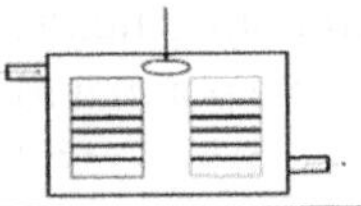	Can be used with surface growing microbial cultures Well established design Drainage	Solid state fermentation *Aspergillus* and soy bean fermentation	[9]
Micro-bioreactor 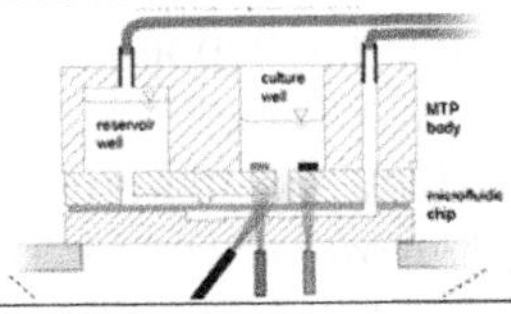	High-throughput capacity Small consumption of media Difficult to control Gradients	Process development and evaluation of any microbial or cell cultures	[13]

Figure 9.1 Comparison of common bioreactor types. Reproduced with permissions from [13], [14], and [16].

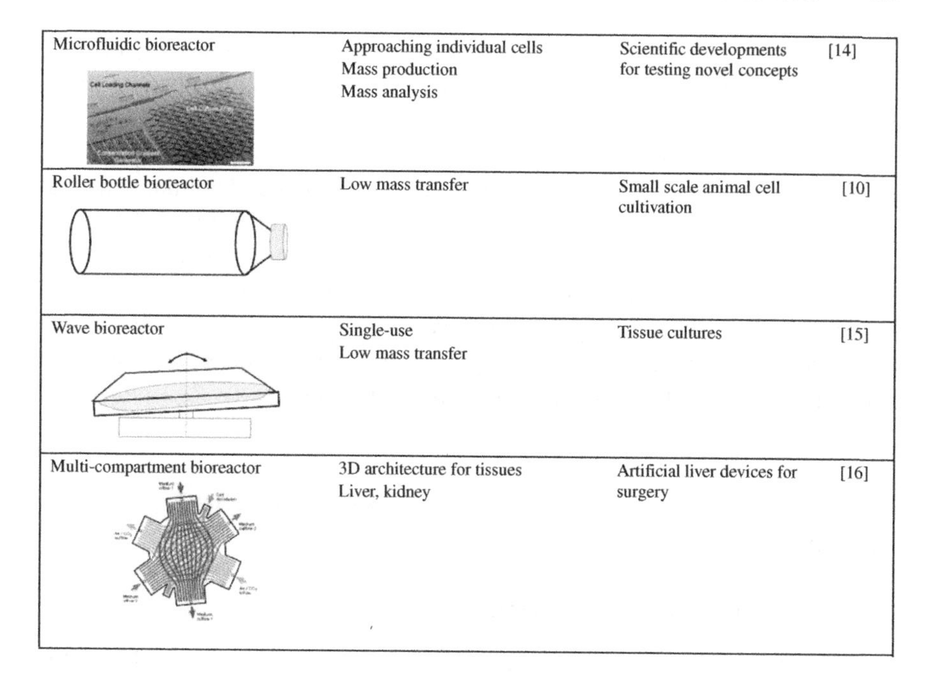

Microfluidic bioreactor	Approaching individual cells Mass production Mass analysis	Scientific developments for testing novel concepts	[14]
Roller bottle bioreactor	Low mass transfer	Small scale animal cell cultivation	[10]
Wave bioreactor	Single-use Low mass transfer	Tissue cultures	[15]
Multi-compartment bioreactor	3D architecture for tissues Liver, kidney	Artificial liver devices for surgery	[16]

Figure 9.1 *(Continued).*

on a platform under gentle motions, and gauges and other permanent instrumentation control the operation. This is definitely a new conceptual design approach. A careful cost analysis for balancing single-use device unit cost versus cost of labor, consumables, and so on, when using other bioreactor designs, is necessary before using them in manufacturing processes. So far, there seems to be a market for this bioreactor type, at least for some applications.

Yet another recent bioreactor system targeted for *tissue cultures* is the rotary microgravity reactor [28]. The cultivation takes place under microgravity created by a centrifugal force created by rotation. Especially, tissue cultures from human cartilage have been successfully applied by NASA in a space station project [29].

Expansion of stem cells in bioreactors is a necessity for further exploitation of these cells in biomedicine and pharmacology. This is probably one of the most demanding bioreactor applications due to the complex differentiation of the stem cells that must be controlled by biological growth and differentiation factors over extended time periods. A good technical solution for this is a challenge for conceptual design [30].

Characteristics and dominating applications of the main bioreactor types differ. Figure 9.1 summarizes these with drawings of the main bioreactor types. As obvious from the overview given here, the technical requirements of the bioreactors are both complex and diverse. Electrical engineering, pneumatics, mechanics, software engineering, pumping technology, mixing, measurement techniques, control methods, and user needs must be integrated into one multifunctional system [31].

The *aim* of this chapter is to try to place these requirements in a conceptual design context. The focus is on production of microbial and animal cells that produce proteins and small biomolecules. All these bioreactors aim for large-scale industrial production of the substances. The scales can be modest, as for production of a particular monoclonal antibody, or as large as 200–1000 m^3 for yeast and bioethanol. By retrospectively analyzing the missions, target specifications, and generated concepts of previous bioreactor designs, a transparent view and understanding of them are gained.

9.2 MISSIONS, USER NEEDS, AND SPECIFICATIONS FOR BIOREACTORS

9.2.1 Design Mission and User Needs

Before considering the needs of the bioreactor users, it is wise to define the mission of the particular bioreactor to be designed and from that detail the needs in relation to that mission [32]. In Table 9.1, the essential needs for a few bioreactor missions are shown.

The table shows how user needs of a bioreactor design vary significantly depending on the intended use. Shall the bioreactor be used for manufacturing, process development, education and training, or any other possible application? The user needs must also clarify what type of process and organisms that are concerned, the scale of production, if high-throughput capacity is required, if a certain product purity must be reached, and so on.

The mission and needs have significant impact on the ensuing design work. For example, a bioreactor having the mission to be used for mammalian cell cultures in a GMP facility allows to limit the design criteria to a lesser number of requirements (i.e., it limits the "design space" or "design freedom"), while a multipurpose bioreactor for education with other cell types and without GMP demands changes the design criteria considerably. The consequences for the manufacture of the bioreactor, the potential market (i.e., number of users of the bioreactor product), and the number of available applications vary.

9.2.2 Target Specifications

The user needs are reformulated as general specifications that the design solution should fulfill. These specifications first put restrictions on the generation of concepts. Primarily, the identified needs come from basic knowledge

TABLE 9.1 Different Bioreactor Missions and User Needs Accompanying the Missions

Mission	User Needs	Typical Bioreactor Design
Production of small molecules (e.g., citric acid)	50–100 m^3 volume Sanitation possibilities Infection-proof	Stirred tank reactor
Production of proteins for industrial enzymes (e.g., subtilisin)	25–50 m^3 volume Sanitation possibilities	Stirred tank reactor
Production of diagnostics (e.g., monoclonal antibody)	Sterilizable 1 m^3	Airlift reactor
Production of biopharmaceutical proteins (e.g., insulin)	1–10 m^3 volume GMP compatibility Process validation prepared Comply with regulatory req.	
Process R&D	Lab-scale apparatuses, sufficient volume for analytics, online instrumentation Could scale-down manufacture	For example, 10 L Applikon *in situ* sterilizable bioreactor
Supply of clinical testing materials (e.g., drug development)	GMP compatibility Process validation prepared	Pilot-scale bioreactor
Cell production for food ingredients (e.g., baker's yeast)	Sanitation, large volume	Large-scale bubble column bioreactor
Organ simulation for clinic (e.g., artificial liver)	GCP compatible	Hollow-fiber multicompartment bioreactor device
Organ simulation for testing chemicals/drugs (e.g., liver model)	Small high-throughput device, GLP	Microbioreactor/ lab-on-chip
Academic education and training	Multipurpose units	Lab-scale bioreactor

of requirements of the reactor and other process equipment. These can be collected from any basic engineering text. This should be complemented with interviews and more systematic enquiries with potential customers and users. Experienced users should be included as well as users who are able to see daily hands-on aspects. People with responsibility for production and economics as well as quality assurance should also be among the interviewed. Table 9.2 lists typical attributes appearing on the specification list. Volumetric capacities, high-throughput, sterility demands, aeration requirements, sample needs, and control aspects should be there. Investment cost, GMP compatibility, and software facilities are few examples of attributes that probably appeared during the need identification but have now been specified.

For the target specification definitions, the list is complemented with target values, that is, the range of values of the attributes associated specification items. The values become the targets for the design generation. If target values are not met, the generated design concept is abandoned. In Table 9.2, the target

TABLE 9.2 Comparison of User Needs for Three Bioreactor Types with Typical Attributes and Metrics

Need → Metrics	Target Values Microbioreactor for Process Development	Target Values Large-Scale Bioreactor for Manufacture	Target Values Stem Cell Expansion Bioreactor Device
Cell/product capacity	10^6–10^{10} cells/mL	10^6–10^{10} cells/mL	10^4–10^6 cells
High-throughput	>1000 parallel reactors	1–5 batches	1–5 batches
Volume	100–500 μL	1–100 m^3	500–1000 mL
Batch time (duration)	1–50 h	10–50 h	>3 weeks
Sterility	High	High	Very high
Aeration	Minor (so far)	High	Modest
Heat transfer efficiency	Modest	Very high	High
Sampling	High	High	High
Shear stress on culture	Low	Low	Very low
Cell types	All	All	hESC, progenitors
Price range	Sensitive	Insensitive	Insensitive
Instrumentation			
Online measurement	Modest–very high	Modest	Rather high
Control instrumentation	Modest	Higher	Higher
GMP requirements	No	Very high	High
Regulatory approval	No	Yes	Yes
Software equipment	Yes	Yes	Yes
Automation	Yes	Yes	Yes
Campaign multiuse	No	Yes	No
Disposables	Yes	No	No

values are exemplified for a number of typical bioreactor attributes that will be further discussed later in this chapter. The general specification attributes are narrowed down with the aid of the target values. These ranges or values become the steering frames for the generation of design concepts of bioreactors. The setting of ranges excludes numerous design concepts. They direct the concept generation to what is essential for the utility of the bioreactor. As seen in the list, different bioreactor types create quite diverse metrics and target values.

For example, a microbioreactor for process development should preferably operate with parallel vessels with small volumes in order to acquire large amount of data and for testing many combinations of strains, media, and operation conditions. The monitoring and control for the microbioreactor units must be acceptable. There are no regulatory demands except that the obtained results may need to be included in files and the analysis methods need to be validated in a transparent way.

On the other hand, if the mission of the bioreactor is a defined production facility or product to be produced in a regulatory approved manufacturing process, for example, for penicillin manufacture as exemplified below, the target specification values will be narrow. The volume and/or capacity is set by the predicted production volume, the sterility demands are defined for a set batch time, the production strain and its requirement of media are known, and so on. The customer can decide exactly what supervision facilities and software process control system that are needed. Thus, a blueprint for a customer-designed single-unit reactor may be generated.

For a bioreactor to be used for stem cell manufacturing, quite different demands are outlined. The radically different methodology for expanding surface growing stem cells, progenitor cells, or derived cardiac cell types generates other design solutions. The boundaries caused by the differentiation and redifferentiation of the stem cells, the possibility to monitor critical biomarkers for characterizing the phenotype, and the sequential addition of controlling factors for stimulating differentiation are examples of limiting conditions for the stem cell bioreactor design.

Target specifications of this help the designers at the outset of the design work. However, additional design tools must be used for detailing the design options.

9.3 ANALYSIS OF SYSTEMS FOR CONVENTIONAL BIOREACTORS

In the following sections, the systems and subsystems necessary for carrying out the transformation process (TrP) in the bioreactor are described from a biomechatronic design perspective. The TrP could be any bioconversion process that is possible to carry out in a submerged microbial or cell culture system, where nutrients are taken up and converted into metabolites or protein products. The Hubka–Eder mapping is used to analyze the interactions between the systems in a generalized way [33,34]. The biological and technical systems entities are described more thoroughly since these are pivotal in a bioreactor that performs biological conversions. To make the discussion of the systems more concrete, two industrially well-known bioreactor systems are described in more detail in order to help the reader recognize the systems in real production systems and to obtain examples of how the biomechatronic design methodology is helpful for analyzing them. The first case is a large-scale bioreactor for the production of penicillin G in a metabolically engineered *Penicillium* strain. The second case is a mammalian cell culture bioreactor producing a recombinant protein using an engineered Chinese ovary hamster (CHO) cell line.

9.3.1 Biological Systems in the Bioreactor

The main purpose of a bioreactor is to control the biological transformations that take place in it – thus, many of the key components of the bioreactor are the functions of the biological systems (ΣBioS) and the transformations their actions achieve.

A common way of describing the transformation process would be to follow the established biochemical engineering approach – to structure the transformation into the biological conversion steps based on metabolic maps and process flow diagrams [1]. This would more or less automatically end up in a description with mass transport and rate constant-based kinetics. This would depend on the environmental state (e.g., temperature- and pressure-dependent constants) and supply of raw materials (media). Also, the TrP of the Hubka–Eder map has an inherent mass balance structure between the inputs and the outputs (see Figure 9.2). The map has defined phases (preparing, executing, and finishing) as in a conventional process flow diagrams, and in the map it is relatively easy to identify those phases where the biological, kinetically controlled transformations take place. The Hubka–Eder map can be adapted to cover typical bioreactor processes such a recombinant protein expression, viral vector production, or stem cell differentiation. A matrix representation of these reactions or actions would then be set up. In principle, this matrix could nicely frame-in the current view on biochemical reaction engineering [7,35]. Biological systems for production of bioproducts such as natural proteins, secondary metabolites, monoclonal antibodies, energy carrier molecules, and so on, could be described with sufficient details. However, this should not be done at this stage in the Hubka–Eder approach.

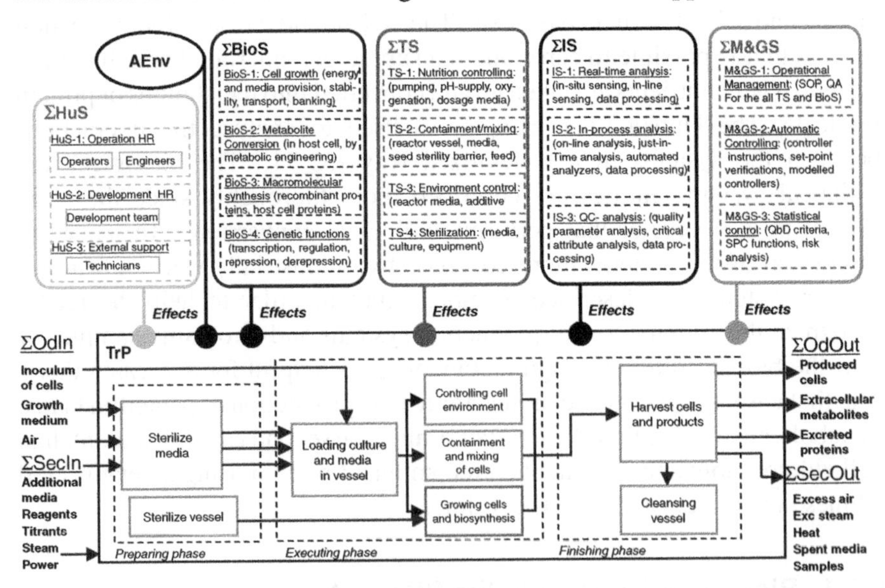

Figure 9.2 *An overview of Hubka–Eder map of a bioreactor. The main systems (ΣBioS, ΣTS, ΣIS, ΣM&GS, and ΣHuS) and first level of subsystems (subsystems[1]) are shown while lower levels of subsystems are mentioned in brackets.*

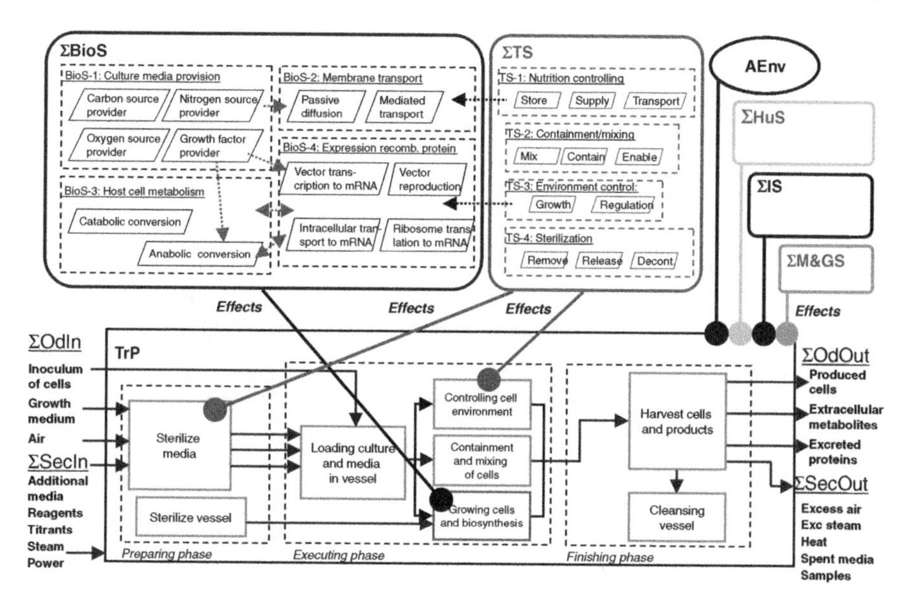

Figure 9.3 *Zoom-in Hubka–Eder map of the biological (ΣBioS) and technical (ΣTS) systems of a bioreactor for recombinant protein production. A selection of subsystems[2] is shown as indicated by numbers in the dotted boxes.*

The advantage of the Hubka–Eder mapping is that it displays all functional actions of the biological systems and from these exhibits the effects the other systems in the map can execute. The current biochemical engineering modeling methods are limited to parts of the TrP without taking this more holistic approach.

In Figure 9.3, the approach is further applied for a well-known biotechnology application: protein production in a recombinant host cell line is exemplified in a zoom-in Hubka–Eder map. The biological systems have in the map been divided into four different biological systems: the culture media system (BioS-1), the transport system of the cells (BioS-2), the host cell metabolism (BioS-3), and the expression system (BioS-4). In Table 9.3, further subdivision is shown where the culture media system is split into different nutrient sources, the membrane transport in passive and mediated transporting, the metabolic conversion into catabolic and anabolic pathways, and the expression system into vector reproduction, vector transcription, and mRNA transportation and translation. For every trained genetic engineer, a further subdivision into subsystems[2] can be done beyond what the table space allows. Also, a subsystem[3] and subsystem[4] can be included, preferably using a software tool to support the structuring of the information. It is noteworthy that at the higher system levels only functions are described, while at the lower levels anatomical structures are introduced, such as a particular nutrient or

TABLE 9.3 Biological Systems, Subsystems and Subsystems[2] (ΣBioS) for a Bioreactor Expressing a Recombinant Protein

Biological System	Subsystem[1]	Subsystem[2]
Culture media	Carbon source provider	CO_2 gas Organic carbohydrates
	Nitrogen source provider	Organic nitrogen-containing compounds Ammonia Nitrogen gas
	Oxygen source provider	Oxygen-containing carbohydrates Oxygen gas
	Growth factor provider	Vitamins Coculture
Membrane transportation for extra and intracellular processes	Diffusion	Membrane phospholipids Cell wall architecture
	Mediated biological transportation	Expression of essential transporters Intracellular transporters
Intracellular metabolism of the host cell	Catabolic conversion	Glycolysis pathway Fatty acid oxidation
	Anabolic conversion	Amino acid pathways Nucleotide pathways
Expression of recombinant protein	Product gene vector transcription to mRNA	Vector structure alternatives Promoter alternatives Induction alternatives
	Product vector reproduction	Vector construct alternatives Vector control
	Intracellular transport of mRNA	Compartment reduction Stabilization of mRNA
	Ribosome translation of mRNA	Generation of rRNA Generation of tRNA Provision of amino acids

biomolecules, for example, 30S ribosome and tRNA-amino acid. When the alternative anatomical units are identified, the analysis is completed and an Anatomical Blueprint can be set up.

9.3.2 Technical Systems

The most essential functions needed in the technical systems (ΣTS) and these functions' interactions with other systems in the Hubka–Eder map of the bioreactor are included in the descriptions in Figures 9.2 and 9.3. Here, the ΣTS have been divided into subsystems for the functions of heat exchange, agitation, pumping (transporting liquids and gases), containment, sterilization

TABLE 9.4 The Technical Systems and Subsystems of a Typical Bioreactor

Technical System	Subsystem	Example of Anatomical Component
Heat transfer	To keep culture at optimal temperature level	
	To sterilize the equipment	
Agitation	Disperse air	Sparger
Transport of media		Bubbling device
	Mix liquid/air	Turbine impeller
		Marine impeller
		Anchor impeller
		Toroid device
		Baffels
	To transport gaseous media	Pressure vessel
		Gas flow system
	To transport liquid media	Displacement pump
		Peristaltic pump
		Syringe pump/device
		Flask transfer device
		Hydrostatic pressure system
Filtration of media	Particle removal	Minifiltration
	Virus removal	Ultrafiltration
	Heat labile molecule removal	Microfiltration
Containment	To contain batches repeatedly	Steel vessel
		Glass vessel
		Teflon vessel
	To contain one batch	Glass jar
		Plastic bag
Sterilization of equipment and media	Sterilization of equipment and media together	*In situ* heat sterilization
		In situ chemical treatment
		In situ radiation sterilization
		Microfiltration
	Gas media sterilization	Microfiltration
		Flush sterilization
Pressure generation	Headspace pressure generation	
	Air gas generation	

(partly overlapping with the previous), chemical-state transformers, and pressure generation. Table 9.4 resolves the map views of the ΣTS in more detailed subsystems and functions, and gives examples of anatomical components.

For example, the heat exchange subsystem needs a subfunction for removal of heat (produced by the culture) that could be cooling coils or a jacket. The heat exchanger subsystem also needs a function for heating up the reactor medium that could be a heat cartridge, hot vapor perfusion, or, again, a heated coil.

When possible we use the same groups of ΣTS for different bioreactor types. Thus, the TS-1 system concerns the function of nutrient handling (supply, store, and transport). On the next system level, this will result in pumping or injection, storage containment of nutrients, and means to contact the nutrients with the cells. For example, CO_2 supply through pH balancing a buffer of gas headspace is a design alternative to be ranked toward the cells' transport functions as described in BioS-2.

The functions of the TS-2 system concern containment and agitation. The TS-2 system should protect of the culture from the environment and, sometimes the opposite, protect the environment and operators from harmful organisms and products. The containment function is also a hub for other parts of the ΣTS.

The mixing function is subdivided into agitation of cells, added media, and supplied gas. Anatomical parts for these subsystems could be turbines, draft tubes, and rotating vessels (cf. tissue culture application below).

Alternative ways of introducing oxygen (e.g., by spargers or silicone tubing) without provoking oxidative or shear stresses on the cells are considered at this stage.

The TS-3 system provides the functions for control of the bioreactor environment. Subsystems involve heat transfer (e.g., by heat exchange, preheated liquid media, and reactor jacket), pH regulation, pO_2 regulation, CO_2 regulation, and pressure regulation and media additives (factors, shear force reducing polymers such as polyethyleneglycol).

The function of the TS-4 system is to provide sterility to the bioreactor. Common operations are *in situ* heating procedures, chemical treatment, and microfiltration. Here, it is suitable to also consider to the less common alternatives such as radiation and toxicant treatment or to introduce disposable bioreactor vessels that revolve the prerequisites for sterilization procedures significantly.

9.3.3 Studying the Interactions of the Systems

The second step in the analysis is to identify the interaction effects and to rank their impact on in the design. Figure 9.4 shows how this is done in a Functions Interaction Matrix (FMI) for all systems in the Hubka–Eder map. In Figure 9.5, the FMI is focused on the biological systems and subsystems shown in Table 9.3. The biological systems can be projected on a systems biology description of the cellular processes. Systems biology models, based on metabolomic, transcriptomic, and fluxomic modeling, can be used as a complementary map for the architecture of the ΣBioS [20–22].

Figure 9.5 goes through the ΣBioS for standard bioreactors producing extracellular metabolites or proteins according to specified requirements. These are applicable to the two cases highlighted below.

The evaluation of the cross-interactions in the matrices may need additional experiments, literature search, or calculations. Several existing process

Main systems	Subsystems	ΣBioS					ΣTS				ΣIS			ΣM&GS			ΣHuS		
		Cell growth	Cell metabolism	Macromol. synthesis	Genetic functions	Cell selection	Media control	Cointainment, mixing	Environmental control	Sterility	Real-time sensing	In-process analysis	QC analysis	Operation management	Controler management	Regulatory management	Operation	Engineering	Support
ΣBioS	Cell growth		5	5	5	1	4	5	4	1	1	4	4	4	1	1	1	1	1
	Cell metabolism	4		4	4	4	2	3	5	5	1	5	5	4	1	3	1	1	1
	Macromol. synthesis	5	4		4	5	4	4	5	5	1	5	5	4	1	3	1	1	1
	Genetic functions	1	5	3		1	1	1	3	3	1	4	4	5	1	4	4	4	2
	Cell selection	4	4	4	4		4	4	4	2	1	4	4	4	1	2	1	1	3
ΣTS	Media control	5	5	5	5	5		5	5	3	1	2	2	3	1	4	5	5	4
	Cointainment and mixing	5	5	5	5	5	5		5	4	1	4	3	3	1	2	3	3	3
	Environmental control	3	4	4	4	4	5	2		3	1	3	3	3	1	2	4	4	3
	Sterility	1	1	1	1	1	1	1	1		4	4	4	4	1	2	4	4	1
ΣIS	Real-time sensing	1	1	1	1	1	1	1	1	1		1	1	5	5	5	5	5	4
	In-process analysis	1	1	1	1	1	1	1	1	4	2		5	5	5	3	4	4	3
	QC analysis	1	1	1	1	1	1	1	1	5	5	5		5	3	2	4	4	3
ΣM&GS	Operation management	4	4	4	4	4	4	4	4	4	4	4	4		4	4	5	5	4
	Controler management	1	1	1	1	1	1	1	1	1	5	1	1	4		4	5	5	4
	Regulatory management	1	1	1	1	1	1	1	1	1	1	1	1	5	5		5	5	5
ΣHuS	Operation	4	4	2	2	5	4	4	1	2	1	5	3	5	3	4		5	3
	Engineering	1	1	1	1	1	3	3	3	3	1	5	4	5	5	4	5		4
	Support	5	5	5	5	5	2	2	4	2	2	2	2	5	5	5	5	5	

Figure 9.4 *A Functions Interaction Matrix that overviews of the main systems (ΣBioS, ΣTS, ΣIS, ΣM&GS, and ΣHuS) that exert effects on the transformation process in a microbial or cell culture bioreactor ((5) is a very strong interaction, (4) a strong, (3) an intermediate, (2) a weak, and (1) a very weak or nonexisting interaction).*

models are here useful as tools for appreciating strengths of interaction effects. An example of such tools is regime analysis methods, as have been successfully applied by Heijnen and van Dijken [36,37] as well as other researchers, to identify bottlenecks of design alternatives. So far, such methods have most of the focus on the biological systems and how to select appropriate strains or cell lines or to adapt media composition, while little concern is dedicated to effects from the technical systems.

The main objective of the Function Interaction Matrix is to compare the interdependence of the biological subsystems functions with the other systems of the bioreactor in a systematic way. This approach could, of course, be applied in any bioreactor design project. In that effort, the subdivision of the ΣBioS into subsystems facilitates the analysis so that effects emanating from the ΣBioS, for example, for recombinant protein expression, different cell lines, plasmid constructs, and media composition, become systematically disentangled.

Biological System			BioS-1 cell growth							BioS-2 metabol conv.		BioS-3 macro-molecular synthesis				BioS-4 genetic functions				
Biological Subsystem[1]			Nutrient provision		Energy supply		Transport ing		Bank ing	Pri ma ry	Se co nd ary	Host proteins		Heterolog ous proteins		DNA tran-scrpition		Repres-sion		De-repre ssio n
Biological Subsystem[2]			Carbon	Factors	Carbon	Oxygen	Nutrients	Gases	Master bank	Phospolipids	Active transp	Glycolyis path	Fatty acid oxid.	Aminoacid path	Nucleotide path	Strructure	Induction	Construct	Vector control	Stabilization
Cell growth	Nutrient provision	Carbon		1	1	1	1	1	1	4	4	5	5	5	5	3	3	3	3	3
		Factors	1		1	1	1	1	1	4	4	5	5	5	5	3	3	3	3	3
	Energy supply	Carbon	1	1		1	1	1	1	4	4	5	5	5	5	3	3	3	3	3
		Oxygen	1	1	1		1	1	1	4	4	5	5	5	5	3	3	3	3	3
	Transport-ing	Nutrients	1	1	1	1		1	1	4	4	5	5	5	5	3	3	3	3	3
		Gases	1	1	1	1	1		1	4	4	5	5	5	5	3	3	3	3	3
	Banki ng	Master bank	1	1	1	1	1	1		4	4	5	5	5	5	3	3	3	3	3
Metab. conv.	Prim ary	Phospholipid	5	4	4	4	4	5	4		5	5	5	5	5	5	5	5	5	5
	Seco nd ary	Active transp	5	5	5	5	5	5	5	4		5	5	5	5	5	5	5	5	5
Macromolecualr synthesis	Host proteins	Glycolysis path	2	5	4	5	2	2	4	1	3		4	4	4	3	3	3	3	3
		Fatty acid oxid	2	2	2	2	2	2	3	5	5	5		4	4	2	2	2	2	2
	Heterolog ous proteins	Amino acid path	2	5	5	5	5	2	4	4	5	5	5		4	5	5	5	5	5
		Nucleotide path	2	5	5	5	5	5	5	5	5	4	3	4		4	4	4	4	4
Genetic functions	DNA tran-scription	Structure	1	2	2	2	2	2	2	2	2	2	2	2	2		5	5	5	5
		Induxtion	1	1	1	1	1	1	1	1	1	5	5	5	5	1		1	1	4
	Repres-sion	Construct	1	1	1	1	1	1	1	1	1	1	1	4	3	5	5		5	5
		Vector control	1	1	1	1	1	1	1	1	1	1	1	1	1	4	4	4		5
	D e.r ep r.	Stabilization	2	1	1	1	1	1	1	1	1	1	1	1	1	1	1	1	1	

***Figure 9.5** A Functions Interaction Matrix that focuses the biological systems ((5) is a very strong interaction, (4) a strong, (3) an intermediate, (2) a weak, and (1) a very weak or nonexisting interaction).*

9.3.4 Penicillin Production in a Metabolically Engineered *Penicillium* strain (Case 1)

This case shows the application of the biomechatronic design methodology to a bioreactor intended for penicillin G production in a metabolically engineered *Penicillium* strain on large manufacturing scale. The biological and technical systems (ΣBioS and ΣTS) will be described here for each of the subsystems shown in Figure 9.2.

The purpose of this descriptive functional analysis is to expose the functions for allowing the subsequent interaction analysis to assess how the systems exert their effects on each other. This will serve the purpose of facilitating the design of the bioreactor. Thus, this should not be mixed up with how the fungus can increase its capacity or quality of penicillin production.

The BioS-1 involves the functions for growth of the *Penicillium* fungus. This includes the growth and production media for fed-batch operation (BioS-1.1). The main nutrient sources for sustaining optimal growth are typically glucose, lactate, ammonia, and oxygen supplemented with a selection of trace elements. The design of the composition of the media could, for example, be supported by using factorial design experiments [38]. The sensitivity of the components for other systems for agitation method, monitoring of rates, and so on are further commented on below.

The BioS-1 also involves the function for transport across the *Penicillium* cell wall and membrane. This is a rate-limiting step. Growth rate cannot exceed this capacity. Its relation to mixing and avoidance of oxygen or nutrient-depleted zones in the vessel is critical.

The BioS-2 concerns the metabolite conversion in the fungal metabolism. The metabolism both for growth and for penicillin production is included. The pathway to penicillin is of primarily concern (BioS-2.2). To enhance the flux of the pathway to penicillin, a metabolic engineering approach can be carried out by elevating the expression of bottleneck enzymes in the pathway. This should be accomplished by the protein production system. Also, strain development and selection is under the scope of the metabolic aspects of BioS-2 for both cell mass growth and metabolite turnover.

The function of BioS-3 is to provide the *Penicillium* proteome. The enzymes involved in the pathway to penicillin G is a separate subsystem (BioS-3.2) here. Improvement of the *Penicillium* strain by metabolically engineering through insertion of a plasmid gene cluster for the pathway enzymes could result in a higher expression level and a higher penicillin flux [20]. The excretion step of penicillin G is also of interest.

Issues of plasmid instability, plasmid number, promoter system, induction method, and selection pressure are also important attributes of BioS-3 subsystems. Stabilizing agents or other means should be considered in the design of the bioreactor.

The BioS-4 system concerns the complex functions of genetic regulation of the filamentous *Penicillium* fungus. The cell cycle is partly well described in general terms. Under consideration here are functions that could interact with the other systems. This should be thoroughly scrutinized even if this seems at this stage unlikely to influence design decisions.

The physiology of the fungus and how it interacts with the agitation of the medium, the delayed diffusion, and the fouling effects on the walls of equipment and detector surfaces are other cross-effects between subsystems, both in the Hubka–Eder maps and the Functions Interaction Matrix (Figure 9.6) [39]. The physiology of the membrane and cell wall structures do also affect extraction equilibrium balance of product molecule in the harvest and recover phase of the TrP. Degradation of the penicillin product and of enhanced enzymes in the penicillin pathway by protease activity is another biological design issue that can be approached by strain selection and genetic engineering.

System			Bios-1 cell growth		Bios-2 metabolite conversion					TS-1 nutrient control			TS-2 containment mixing			TS-3 environ-mental control		TS-4 sterility		
Subsystem[1]			Med. sup-ply	Transport	House-hold metabol.		Pen G pathway		Excretion	Feed	pH	O2	Tank	Feed	Zones	Chem.	Phys.	Tank	Gas	Media
Subsystem[2]			Carbon supply	Transporter activity	Metabolic activity	Flux	Mutation	Feedback block	Side reaction	Pumping	Titration	Pumping	Materials	Pumping media	Impeller numbers	pH	Temperature	Steaming	Microfiltering	Steaming
BioS-1 growth	Med. supply	Carbon supply		5	4	5	1	4	5	1	3	3	1	4	3	4	2	1	2	5
	Transport	Transporter activity	1		5	5	5	5	5	5	5	5	5	5	4	4	4	4	4	4
Bios-2 metabolic conversion	House-hold metabol.	Metabolic activity	1	1		2	2	4	5	2	3	1	1	1	1	3	3	3	3	3
		Flux	3	3	3		3	3	3	3	5	3	3	2	3	3	3	3	3	1
	Pen G pathway	Mutation	2	2	2	2		4	4	4	5	4	4	2	3	3	3	3	3	1
		Feedback block	1	1	1	1	1		4	4	5	3	3	2	2	2	2	1	3	1
	Excretion	Side reaction	2	2	2	2	3	3		2	2	2	1	2	3	3	3	3	3	1
TS-1 nutrient	Feed	Pumping	4	4	4	4	4	4	4		3	3	3	3	3	3	3	3	4	1
	pH	Titration	5	5	5	5	5	5	5	4		4	4	4	4	5	5	4	3	1
	O2	Pumping	1	2	2	2	2	2	2	2	2		3	4	4	4	4	4	5	1
TS-2 containment	Tank	Materials	4	4	4	4	4	4	4	4	4	4		3	3	3	3	3	2	1
	Feed	Pumping media	1	1	3	3	3	3	3	3	3	3	3		1	1	1	1	1	1
	Zones	Impeller numbers	5	5	5	5	5	5	5	5	5	4	4	4		5	5	5	3	1
TS-3 environ.	Chem.	pH	4	4	4	4	3	3	3	2	2	1	5	5	5		5	5	3	3
	Phys.	Temperature	3	3	3	3	3	3	3	3	3	4	4	4	4	4		3	3	3
TS-4 sterility	Tank	Steaming	3	3	1	5	5	5	5	2	2	3	3	2	4	2	5		1	3
	Gas	Microfiltration	5	5	5	5	5	1	2	1	2	3	3	2	4	4	3	3		3
	Media	Steaming	2	1	3	5	5	4	4	4	4	4	5	4	4	3	3	3	3	

Figure 9.6 *A Functions Interaction Matrix for the biological (ΣBioS) and technical (ΣTS) subsystems in a bioreactor producing penicillin G using a Penicillium culture ((5) is a very strong interaction, (4) a strong, (3) an intermediate, (2) a weak, and (1) a very weak or nonexisting interaction).*

Nutrients for industrial fungus fermentation are often semipure, less defined media. In spite of being a pharmaceutical process, they are still approved due to their long use without observed negative side effects. Thus, TS-1 functions (supply, store, and transport) interact with BioS-1 (culture media) functions with regard to the carbon and nitrogen sources. Especially, natural hydrolyzates are viscous and pumps designed for this purpose must be chosen. The media are also more vulnerable for infection than other media alternatives and require considerations concerning sterilization procedures, that is, an interaction with TS-4 functions. Conversely, other media should be considered for BioS-1 functions.

Supply of oxygen in the viscous fermentation broth, especially at the end of the cultivation when cell density is high, puts high demands on the capacity of impellers or other alternatives of the agitation function.

For fed-batch bioreactors and media tanks, the containment function, TS-2, is typically carried out in vessels up to 100 m^3. When preparing feeding during fed-batch phase of more than one influent flows require efficient sterilization capacity (TS-4). The fed-batch time is more extended than for many other fermentation processes, which also puts a burden on the sterility function.

Cleansing of the containment vessel of other parts of containment become more demanding with clogging and fouling media and fungal suspension – this can also be included in the sterilization and sanitation procedures.

The environment control functions, TS-3, are also affected by the state of the extracellular system in the bioreactor for the same reasons. More gradients are created both due to large vessel size and due to viscous media. If more parts are installed inside the vessel, stagnant zones with depletion of the media may become abundant.

Means to optimize, monitor, and control the bioreactor are generated in the ΣTS, ΣIS, and ΣM&GS, which are further discussed in the following sections.

Again, this analysis of functions in a bioreactor design for a particular application, large-scale penicillin manufacture, has the sole purpose of sustaining the efficiency of the biomechatronic design approach when designing a bioreactor system, and is not a way to analysis how to develop the antibiotic bioprocess. That can also be done with the support of the biomechatronic theory, but would then require an approach for structuring the analysis other than above.

9.3.5 A Bioreactor System Producing a Recombinant Protein in CHO Cell Culture (Case 2)

In this case, the biomechatronic design methodology is applied to a bioreactor for production of a recombinant protein by a CHO cell culture. This means in a number of differences in needs and conditions for the design. Still, the same main systems as in the previous case are appropriate to analysis. But these need to be supported by additional subsystem entities that represent the specific functions of the mammalian cell physiology such as glycosylation, chaperonin expression, apoptotic regulation, and surface attachment. Also, the technical design of a bioreactor for mammalian cells requires special attention in a number of ways.

For a mammalian cell culture, the functions of BioS-1, cell growth, have a greater complexity. Especially, the BioS-1.2, that is, the media supply function, concerns here the media of a larger number of components. It could include, for example, two different nutrient compositions, one for CHO cell growth and another for protein production. It could be growth in serum or serum-free media. Since the media are significantly more complex than the media for antibiotic production in fungi, they are supplied with expensive physiology stimulating factors, something that significantly increases the media cost. Consequently, this emphasizes its importance in designing.

The BioS-3, that is, the macromolecular synthesis function, is greatly affected by a more complex proteome setup. For example, transporter proteins located in external and internal membranes that take up the media components, transfer metabolites and proteins, and excrete the product protein constitute a huge complex network. The BioS-3 for the metabolic machinery of the CHO cell is complex as well and includes only those cell functions associated with cell propagation. The posttranslational functions are here excluded, but instead found together with the other functions involved in protein formation.

The functions for the expression machinery of the recombinant protein, BioS-3.2, include the transfection of the cell line with a product protein-expressing vector, its control functions, the genome integration mechanisms and vector stabilization, the transcription and translation mechanisms, and the glycosylation of the peptide in the endoplasmic reticulum and Golgi compartments.

There are a number of design criteria that show dependence on other subsystems of the bioreactor design. The interplay with shear forces is created by the ΣTS, the cell membrane properties, the mechanical stress, the oxygen stress, and their propagation to other cell functions such as cell division rate, onset of apoptosis, and cellolytic effects, and activation of hsp gene responses. Additional functions possible to consider are stability of vector integration in CHO cells, secretion from CHO cell of the recombinant glycoprotein, chaperone functions, phosphorylation, and compartmentalization effects on the CHO cells.

A general consideration for all mammalian cell cultures is the sensitivity to shear stress, low oxygen transfer, and lower cell densities. These properties of mammalian cells have been main issues in the development of cell culture bioreactors. Thus, designing of theTS-1 functions to reduce harmful effects on the cells has created solutions such as low shear-stress impellers (e.g., marine impellers), oxygenation through porous silicone tubing, introduction of protective polymers in the culture media, the use of microcarriers and airlift oxygenation of these, and many other more or less intricate solutions. It is obvious from this that the transformation functions of media, including gases, require special attention.

The long culture times, sometimes up to 30 days (e.g., human Factor VIII with recombinant CHO cells) put higher demands on containment. This is a priority since the economic consequence of lost batches due to infections is huge. Investment in a more elaborate containment design pays off. The containment functions on an anatomical level become matters of generating alternatives of bearings, gaskets, pipe steaming, polishing of surfaces, and so on.

The sensitivity of the CHO cells to various forms of biological stresses in a wider context needs to be addressed during the design of all ΣTS. But it is in particular in the functions executed by TS-4 where the extracellular environment of the bioreactor can be designed in a better way to cope with biological stress effects. Thus, the interconnections between the ΣBioS and the TS-3 should be scrutinized thoroughly.

Mainly three interactions require special attention: (1) effects on the substrate uptake, (2) effects on the metabolism of the cells, (3) and effects on the recombinant protein product. First, the temperature gradients of the microenvironment of the cells could potentially influence transport rates and membrane fluidity and by that facilitate or impede the exchange of substrates, factors, and so on. Depletion may elicit stressor molecules. Examples are heat shock proteins (hsp 60 and hsp 70), chaperonins, and transmembrane glucotransferases [40].

Second, the apoptotic effects on the CHO cell have external elicitation routes. Can these be alleviated by tuning the microenvironment? Can injections of factors be controlled? These questions touch upon the state of the art of cell biology.

Third, can the conditions in the microenvironment be controlled so that proteases degrade the recombinant product inside or outside the cell (e.g., due to lytic release)?

The control possibilities of the microenvironment by ΣTS-3 are limited to pH and pO_2, temperature, possibly pressing, shear forces, pCO_2, and a few measurable substances. To realize these control "loops," ΣIS functions (see below) must be added, some today nonexisting but subject for intensive research.

The Functions Interaction Matrix in Figure 9.7 ranks the strengths of interactions between the biological and the technical systems of the bioreactor. These interactions require a thorough research of data and information, but are still a time-saving effort compared to experimental trials with a low growth rate CHO cell culture.

The two cases show the value of applying the biomechatronic methodology directly to the intended applications of the bioreactor product. By that, the user needs become even more highlighted in the design, the added complexity of the ΣBioS better integrated, and the potential problems the users will meet anticipated, relieving the product support functions.

9.3.6 Information Systems

The exchange of information between the biological and the technical systems in the bioreactor and the bioreactor management goes through the information systems (ΣIS). Their functions are to acquire and interpret signals and present these signals in forms that are useful for monitoring and controlling the transformation in the bioreactor toward set goals [41].

System			Bios-2 Membr. transp.		Bios-4 Expression of recombinant protein					TS-1 Nutrient control			TS-2 Containment mixing			TS-3 Environ mental control		TS-4 Sterilization		
	Subsystem[1]		Dif fus ion	Me dia ted	Vector transcript		Vector reprod		m RNA	Store	Su ppl y	Tran spor t	Mix	Harv -est	Cont ain	Grot h	Regul at	Rele ase	Rem ove	Rem oe imp
		Subsystem[2]	Phospolipids	Act. transporter	Structure	Induction	Construct	Vector control	Stabilization	Tank vessels	Injection	Pumping	Turbine	Harvest tank	Steel vessel	Feeding regime	Injection base	Detergent	Heating	UV radiation
BioS-2 Membr.	Diffusion	Phospholipids		5	4	3	3	3	2	2	3	2	2	2	2	3	3	3	3	2
	Mediated	Active transporter	5		4	4	4	3	5	1	1	1	1	1	1	1	1	1	1	1
BioS-4 Expr. of recomb. protein	Vector transcript	Structure	2	2		4	4	4	4	1	1	1	1	1	1	1	1	1	1	1
		Induction	3	4	4		5	5	5	1	5	1	1	4	2	3	4	1	3	1
	Vector reproduc	Construct	3	2	5	5		5	5	1	5	1	1	4	2	1	4	1	3	1
		Vector control	3	2	4	4	4		5	1	1	1	1	4	1	1	4	1	3	1
	mRNA	Stabilization	1	1	1	1	1	1		1	1	1	2	2	2	2	2	2	2	2
TS-1 nutrient contr.	Store	Tank vessel	3	3	1	1	1	1	1		3	4	5	1	1	4	4	5	4	4
	Supply	Injection	3	2	1	4	4	4	1	2		3	5	1	1	4	4	5	4	4
	Transport	Pumping	3	3	1	1	1	1	1	2	3		5	1	1	4	4	5	4	4
TS-2 containment	Mix	Turbine	5	4	1	5	5	5	1	5	4	4		1	1	3	3	5	4	4
	Harvest	Harvest tank	1	1	1	2	2	2	1	4	4	4	5		1	1	1	5	4	4
	Containm	Steel vessel	3	3	1	2	2	2	1	5	4	4	5	1		1	1	5	4	4
TS-3 environ	Growth	Feeding regime	3	3	5	4	4	4	4	5	4	5	1	1	1		1	2	2	2
	Regulate	Injection base	4	4	5	5	5	5	4	4	4	5	1	1	1	3		2	2	2
TS-4 sterilization	Release	Detergent	5	5	3	3	3	3	3	4	3	4	3	3	1	4	4		4	4
	Remove mo	Heating	5	5	3	3	3	3	3	5	3	4	3	3	1	4	4	4		4
	Remove impurity	UV radiation	3	3	3	3	3	3	3	4	3	2	1	3	1	2	2	4	4	

Figure 9.7 *A Functions Interaction Matrix for biological (ΣBioS) and technical (ΣTS) subsystems for a bioreactor expressing a recombinant protein in a CHO cell culture ((5) is a very strong interaction, (4) a strong, (3) an intermediate, (2) a weak, and (1) a very weak or nonexisting interaction).*

We regard here as ΣIS those systems that provide the functions required for processing the signals. Thus, the technical means that generate the information signals are instead inclusive of the ΣTS. For example, the technical functions of an analytical instrument (e.g., an HPLC) for sample injection, pumping, and detecting are not considered in the ΣIS.

As seen in Figure 9.4, the ΣIS are divided into systems for real-time analysis, IS-1, in-process analysis, IS-2, and quality control (QC) analysis, IS-3.

IS-1 includes such information systems as are able to monitor the TrP in the bioreactor in real time. The purpose is to act on the information in due time. Typically, IS-1 is accomplished by online sensors and devices that can be placed *in situ* in the culture or directly in the technical component (e.g., a valve or a motor).

IS-1 could be the information acquired from such sensors and measurement devices commercial bioreactors are equipped with as a standard setup (e.g., sterilizable pH – and dissolved oxygen electrodes, pressure gauges, gas flow meters, temperature sensors for the reactor vessel, and temperature sensors for sterilization monitoring). IS-1 could also be acquired from new solutions of instrumentation or by combination of signals. One example of the latter is software sensors [42].

Technically, standard setups and new solutions belong to subsystems of the ΣTS. When generating information, either they interact with other technical subsystems (e.g., valve, pumps, and heat exchangers) or they interact directly with subsystems of ΣBioS (e.g., microbial cells, metabolites, and proteins in the culture medium).

IS-2 includes information acquired from in-process analysis. Here, we mainly refer to offline measurements carried out in the bioreactor by operators or possibly by automated analyzers. This information is consequently a result of interaction with people that should follow certain analytical protocols or by machines programmed for certain analytical procedures. Both are external from the physical bioreactor but could, after a time delay, result in interactions with the bioreactor. Examples are dry weight samples from the bioreactor measured by a technician in a spectrophotometer, or testing of glucose in the medium using a dipstick.

The IS-3 includes information acquired from the QC laboratory. Typically, this information is the result of analyses carried out with validated methods required by regulatory demands. It could also be samples sent to the research laboratory where advanced analysis methods are applied. The IS-3-derived information has one thing in common with IS-2 that the information is generated outside the bioreactor by other instruments and by humans (technicians and researchers) in lab facilities separately from the bioreactor. The information may interact with the ΣBioS and ΣTS but rarely in real time. Instead, the information is used by the ΣM&GS for setting limits, establishing procedures, and making decisions. Examples are gene sequence analysis, GC-MS of glycosylated product forms, SDS-PAGE, immunoblotting methods, and so on.

In the IS-1, IS-2, and IS-3 systems, signal processing, data processing, statistics, and modeling methods are useful and become subfunctions of the ΣIS. The information systems could benefit from interacting in between.

It is motivated to thoroughly consider the needs of the information. The required information should be listed and characterized. In what ranges should information be acquired, how fast, and with what precision and accuracy? On the basis of these specifications should the information method be selected. In Table 9.5, examples of information systems are given on two subsystem levels. In the second level, analytical methods and instrumentations are mentioned. A priority of information needs is of value. Analytical sciences can often provide alternative methods, but these must be evaluated toward needs and demands.

TABLE 9.5 Examples of Information Systems Useful in a Bioreactor Design

Information System	Subsystem[1]	Subsystem[2]
Real-time information	Providing information for physical parameters	Temperature sensors Pressure sensors
	Providing information for physical parameters	pH sensors pO_2 sensors
	Providing information for biological parameters	Biomass sensors NADH sensors
Delayed advanced at-line information	Extracellular state information generation	Online HPLC Online enzyme electrode device (YSI)
	Intracellular state information generation	NMR online Imaging microscopy
QC analysis requirements	Existing analytics for generating regulatory documentation	Genome characterization Protein identity verification
	Foreseen methodology for new guidelines	Chemometrics methods Microarray analysis

In particular, the content and character of BioS-1, BioS-2, and BioS-3 have decisive effects on the requirements of ΣIS. For example, bioreactor with culture producing a metabolite or protein would require information on product yield and biomass growth online, by-product formation levels, consumption of nutrients such as glucose and lactose, uptake of ammonia, and so on. The just-in-time parameter is related to the need of monitoring the conversion and the course of growth and product formation and by that furnish the possibilities to take adequate actions. With a successful implementation of the IS-1, this would be possible, but this implementation highly depends on which biological systems are in use in the bioreactor.

One should also evaluate the consequences of information deficiency. This provides another perspective of the effects exerted by ΣBS on ΣIS. The type of cells, their production pattern, growth, metabolic rates, and so on decide if the suggested IS alternative is useful or not. This may be an important design constraint that should be uncovered early and if possible solved with another choice of information device (e.g., a faster HPLC column, installation of online equipment). Such lack of information could lead to a substantial amount of discarded batch because information is delivered too late to initiate compensatory action. This notion may lead to additional need for development of such devices, or, it may help in identifying problems not possible to solve with available techniques. The latter could, for example, result in choosing another organism.

In addition, it is highly important to identify interactions and effects of the performance of the information-generating devices. The required performance can be preferably included in Table 9.5 as operation ranges for the information devices.

Process analytical technology (PAT) has highlighted biopharmaceutical processing where the bioreactor is an integral part. In PAT, a number of ΣIS issues are thoroughly attended [43]. For example, real-time monitoring and data analysis are main concerns of PAT. We have devoted Chapter 13 entirely to PAT and its role in manufacturing processes, with a wider perspective. Clearly, PAT has a role in the basic design concepts of bioreactors.

9.3.7 Management and Goal Systems

The processed information from ΣIS is handled by the management and goal systems (ΣM&GS). It is possible to distinguish subsystems that in a direct way act on the information by furthering instructions to the ΣTS for correction states or conditions related to the ΣBioS. This can be done as instructed in the Standard Operation Procedures (SOPs) where goal values might be set. Typically, it is the operator at the plant that effectuates this instruction initiated by the information delivered. Here, we refer this as a function of M&GS-1.

The information may also be handled in an automated fashion, for example, with a tuned feedback controller with a given set-point. The set-point could be a result of previously established optimal operation conditions for the process. It could also be set by ranges established from quality-by-design (QbD) investigations [44] (see also Chapter 13). We refer to these functions as M&GS-2.

ΣM&GS have also functions independent of the ΣIS. These include the laws and regulations that must be met for a bioreactor system. Here, we group these into M&GS-3. For example, the regulatory authorities have strict regulatory frameworks a bioreactor operating in a manufacturing process must comply with. The Good Manufacturing Practice regulations should be applied to the bioreactor system if it is producing pharmaceuticals. Other quality systems could apply to other products. Environmental protection laws are other examples. Several of the constraints have to be converted into operational procedures that form the instructions of the SOPs of M&GS-1.

Recent initiative from the authorities concerning PAT and QbD should also be implemented by M&GS-3 [43,44]. In the QbD guidelines, the design space and control space of the process to be carried out are established on the basis of experiment studies. The PAT guidelines strongly recommend the use of online monitoring and control, thus directly applicable to M&GS-2, IS-1, and ΣTS (see also Chapter 13).

From these various functions in ΣM&GS, three levels of activities can be distinguished:

- Basic management of the operations to reach acceptable performance of the bioreactor.

- Application of automated control methods based on process information.
- Integration of the operation into regulatory procedures from the plant management, the process engineers (process R&D), according to regulatory guidelines.

The design solutions in ΣBioS and ΣTS depend on the constraints and possibilities that ΣM&GS can cope with. Thus, it is decisive to assess these interactions and effects or at least to have in mind what is needed for these aspects when selecting from generated design concepts.

In most cases, the prerequisites are that the basic management system for operating the bioreactor may follow the SOP protocol according to GMP guidelines, or any other manufacturing protocols. The effects of these plant management procedures on ΣTS, ΣIS, and ΣHuS should be tabulated actions and checkpoints. This is congruent with most quality system and risk management procedures as described in several industrial guidelines, in particular for pharmaceutical manufacture.

The basic management is preferably carried out at the bioreactor via an operator–computer interface. The management structure and checkpoints can be built-in into the interface view. Of the more obvious interactions that are displayed, this interface becomes to some extent a mirror of the TrP in the Hubka–Eder map. In other words, the operator interface should prioritize the most decisive interaction toward ΣTS and ΣBioS via ΣIS. The ΣM&GS should then provide the means to effectuate the control and steering of the bioreactor. Considering a single-unit bioreactor, the situation is relatively easy to overview, while a bioreactor assembly in a plant magnifies the demands of statistical process control (SPC) procedures. The limits for the SPC are set beforehand, that is, as a result of process R&D or goals established by the persons responsible for quality. By this, the goals are set for the operation of the bioreactor. Examples of limits required to be set are temperature at sterilization, noise levels of electrodes before replacement, overpressure alarms, rate of impeller, warning temperatures for the culture, and pO_2 electrode limits for feeding or mixing rate.

An integral part of the management is to take actions. If the actions can be automated by feedback or feedforward control repeatability and consistency of the actions can be expected. The control rules to be followed primarily include basic control methodology, such as on–off controller, P, PI, or PID controllers, or sequence controllers. Set-points and time schedules are normally a result of early development of the controller or more or less elaborate tuning of them.

Sterilization schedules are based on real-time temperature and pressure sensors (i.e., IS-1 and IS-2) and valve and steam generation from TS-4 (cf. Figure 9.3). The effects of signal response times, noise, and accuracy are interactions that have direct effects on the design of the controller. Choice and spending on sensor quality devices become ranking criteria. The impact and requirement on quality of sterilization is a design decision to be taken.

9.3.8 Human Systems

The human systems (ΣHuS) involved for carrying out the TrP can be divided into those furnishing the operation functions of the bioreactor, HuS-1, those providing or using the bioreactor for development functions, HuS-2, and those responsible for the external technical support functions, HuS-3. It is easy to identify groups of professionals in the ΣHuS (technicians, plant engineers, developers, managers, and service technicians). However, at this stage, it is more interesting to consider the functions these professionals furnish and how those functions interact with other systems.

Clearly, the operation of the bioreactor and the HuS-1 interaction with ΣTS are highly important. The conditions vary much from an industrial manufacturing environment to a research laboratory. The training and experience of the operators must be considered. The ease with which the devices are operated, the access to information and data, and the expected actions to be taken by operators and plant engineers determine several of the design criteria placed on the systems. The choice of alternative within ΣTS should consequently be done with considerations of how ΣHuS will interact with them.

As mentioned above, there are many interactions between the ΣHuS and the ΣM&GS concerning information, control, and regulatory issues. The ΣHuS normally initiate or supervise the issues under ΣM&GS. For example, the plant engineers will take the responsibility of implementing and verifying the rules and guidelines, while the development engineers will optimize parameters and investigate suitable control space. A sometimes neglected function is support and maintenance by human systems. If these are not well trimmed, operation will stall in the long-term perspective.

9.3.9 Active Environment

Unexpected effects exerted by the ambient surrounding on the subsystems and TrP in the bioreactor are recurrently experienced by every process operator with great distress. The term active environment (AEnv) covers everything that can happen and cannot be properly explained and foreseen. The so-called "biological variation," usually meant as unreproducible or unanticipated variations in the culture, falls within the category of effects exerted by the active environment. Other effects are involuntary infections, impurities of media, technical malfunctions of equipment, and so on. No doubt, this category of effects is a very important design factor to consider. Its negation by cautious actions that are built into the design could be a potential quality improvement of the bioreactor product.

One should be observant of the fact that the AEnv effects are also exerted on embedded functions of ΣTS and ΣIS and the propagation of these to ΣM&GS. For example, the effects on response times of all the systems could have considerable consequences. Growth from an infection on pH membrane or in ultrafiltration units may retard the expected rate of response of purification time. Fouling on mixers may change shear stress effects. The burden on sterilization equipment is another effect that is difficult to anticipate. Others are heat variation effects on HPLC and LC-MS equipment.

It may be difficult to realize all possible effects of temperature variation in the ambient air or variation in the cooling water line. For example, effects on such devices or parts that are not temperature controlled could be easily disregarded. Effects on DO-electrode operation and "filtration constants" for air filters in the bioreactor outlet are easy to forget. Another example seldom controlled is fouling on heat exchangers affecting the heat conduction number.

More long-term effects caused by ambient could be corrosion effects on metal parts of the equipment. The risk estimates for such effects to interfere in the performance of the bioreactor are of particular value to include in the design decisions.

9.4 NOVEL BIOREACTOR DESIGNS

9.4.1 Microbioreactors

Scaling down of microbioreactors has been accomplished down to working volumes below 50 μL. The main reasons behind this development of microbioreactors are the desire to carry out high-throughput testing of chemicals and to facilitate bioprocess development by faster and more efficient testing of strains and media components [45]. Other microbioreactor applications considered are small-scale production and pharmaceutical development. When including lab-on-a-chip applications in the microbioreactor concept, which is definitely motivated, these possibilities become even more pronounced [46].

The design of miniaturized bioreactors makes an appropriate rearrangement of the Hubka–Eder map. The purpose is to highlight critical obstacles necessary to surpass (Figure 9.8). The physiochemical effects on cells and microorganisms caused by the smaller dimensions must be understood and handled with caution. In particular, gradients and interactions with wall materials become sensitive design parameters. Therefore,

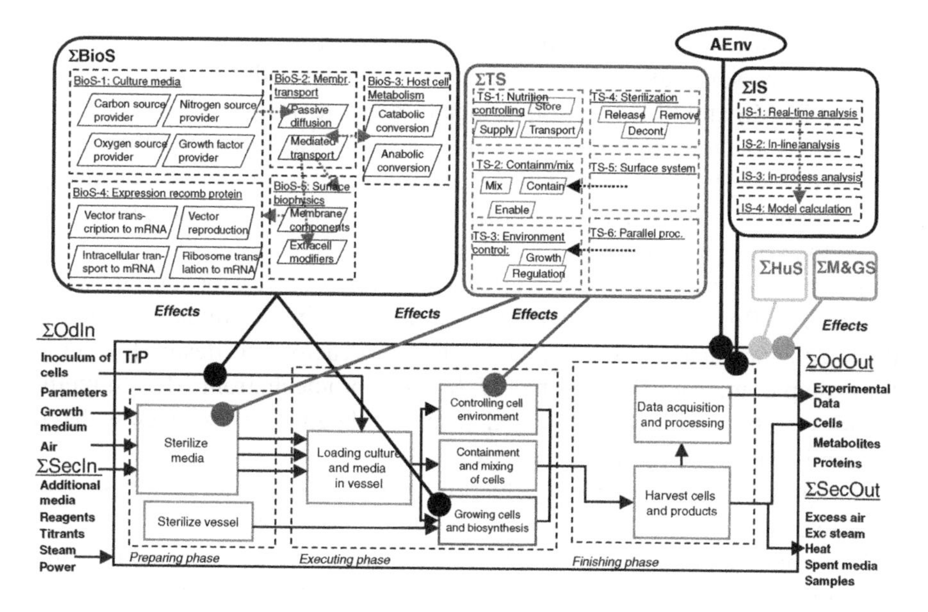

Figure 9.8 *A zoom-in Hubka–Eder map of a microbioreactor system.*

a few additional subsystems[1] are introduced for microbioreactors in the Hubka–Eder map:

- A surface-related function in ΣTS for a more detailed analysis of the alternative concepts related to biophysics of materials (TS-5).
- A technical function for running operations in parallel to increase high-throughput is addressed by a separate (TS-6).
- A biological function of surface interaction due to the cells' adhesion properties (BioS-5).
- An information function related to the analysis of data generated in the microbioreactor based on its properties (IS-4).

These subfunctions could also be introduced for other bioreactors, but the influences on the design of these are so pronounced for microbioreactors that it is unavoidable here.

Figure 9.7 zooms in the new subfunctions and adds a few subsystem[2] alternatives. The BioS-1 is related to the design of the biological membranes by utilizing specific biosurface properties of the cells, mainly active membrane biopolymers (e.g., adhesins, surface binding glycoproteins, and hydrophobicity properties of the lipid composition of the cell surface). The BioS-2 function is related to the external biochemical means to change the surface properties of the cells.

TS-5 focuses on the properties of construction materials of the microbioreactor vessels/wells and other physical parts. Typically, plastic materials such as PDMS and PMMA or other means to create or fabricate surfaces are considered. If the design solution causes strong shear forces or other stress effects due to the containment and oxygen transfer, these effects are a result of interactions of TS-2 and TS-3 with BioS and thus included in the previously mentioned design aspects.

The most critical interaction effects in a microbioreactor are shown in Figure 9.8. Recreating efficient mixing and negating gradient effects and depleted zones in a small volume of a microbioreactor puts design limits on, for example, microwell solutions [47,48]. The biomechatronic design methodology steps should be rehearsed from the beginning and reconsidered (which is partly a reiteration exercise).

In Figure 9.9, an Anatomical Blueprint of the Hubka–Eder map is shown. Concrete components for use in microsystems are inserted.

The challenges of pumping and sensing on microscale are significant. Inventive design solutions are required, as already suggested, for oxygenation with gas-permeable membranes [26,49], spectrometry fiber sensors [50], fluid dynamic analysis [27], and mass transfer [25].

Microsensors and micromechanical devices for mixing and dispersion permit miniaturization of microbioreactors down to very small volumes. Sometimes, the miniaturization does not fulfill an expressed need – the cost reduction due to less media consumption is negligible and the supply of test strains is not a limitation. The volume required for sampling may make further volume reduction impractical or lead to imprecise analysis.

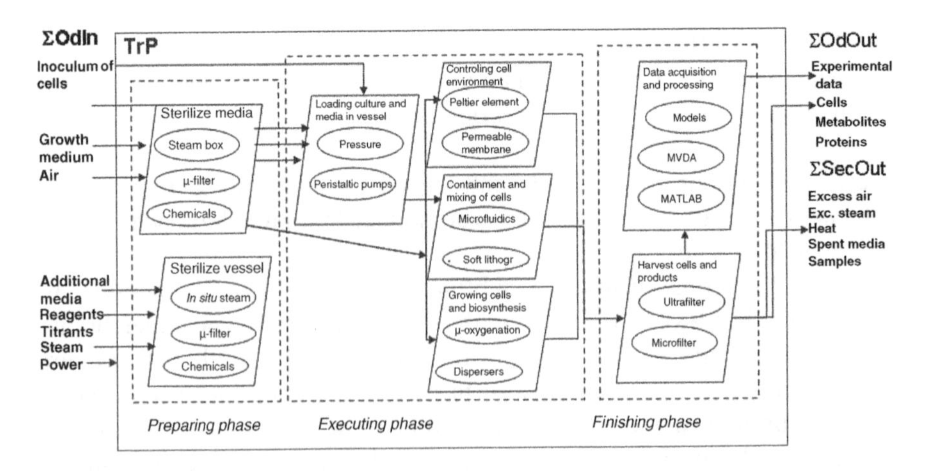

Figure 9.9 *An Anatomical Blueprint of a microbioreactor design intended for data analysis.*

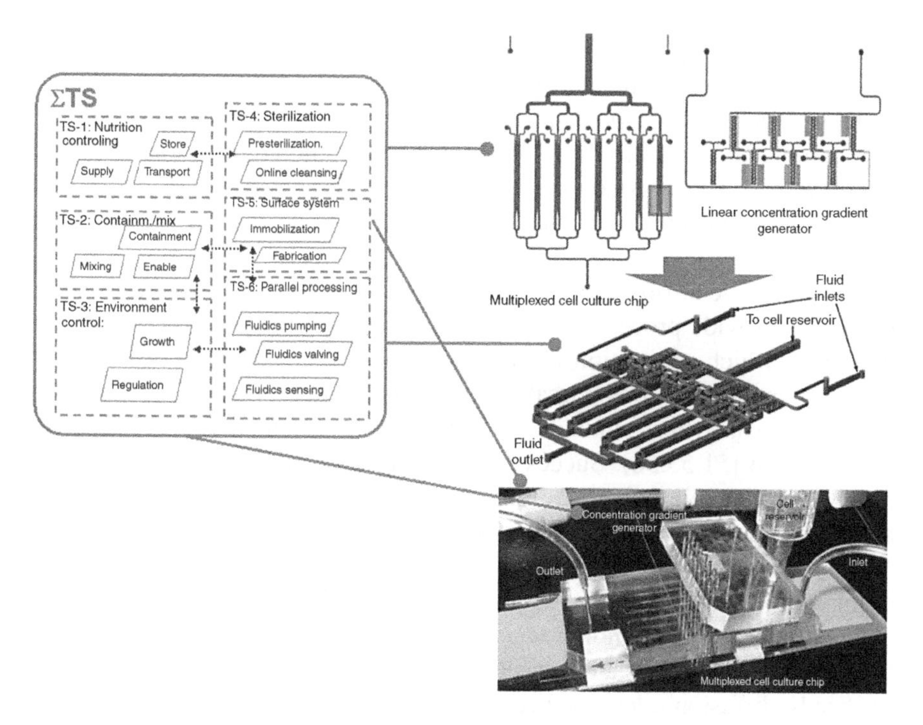

Figure 9.10 *A lab-on-a-chip microbioreactor device with systems, subsystems[1] and subsystems[2] identified. Adapted and reproduced with permission from [52].*

Lab-on-a-chip microbioreactors refer to integration of the bioreactor components on one chip. Lithography techniques applied using, for example, silicon or soft organic polymers, allow miniaturization even to dimensions of single cells and mass fabrication (e.g., [14]).

The utility of the extreme downsizing may sometimes be questioned. However, three unique applications are possible or envisaged: using the device as a research tool for metabolic studies, using the device as a microproduction unit of biomolecules, and using the device for artificial reproduction [46]. For example, the *in vitro* fertilization of oocytes can be performed in a microfluidic system [51].

Also, for the lab-on-a-chip applications, the Hubka–Eder subsystems can be identified as illustrated in Figure 9.10 for a microfluidic device with hepatocyte for *in vitro* testing of drugs [52].

9.4.2 Bioreactors with Immobilized Cells

A popular approach for stable operation of bioreactors is to immobilize the cells [51,53–55]. The immobilization protects the cells and allows their reuse. When the cells excrete molecular products, the immobilized cells are retained and the product separated from the immobilized cell particles. This becomes an efficient and economical process setup for product recovery.

The immobilization procedure may, however, damage the cells during fabrication and/or disturb the cells because of the altered microenvironment. However, biophysical conditions inside the immobilized cell particle can also be improved and by that enhance intracellular processes leading to better productivity.

The extensive efforts devoted to immobilization technology of cells since the 1980s provide process designing with technical information and know-how with which generation of bioreactor design can take advantage. For example, most of the theoretical and technical requirements of the immobilized system have been studied thoroughly and several models are reported in the literature [51,53,56]. Successful applications of the technology are immobilized yeast cells in hydrogels and the industrial production of monoclonal antibodies in microcarrier cultures (e.g., [57,58]).

Thus, to convert the experience and know-how of immobilization technology into mechatronic design principles is relatively easy. An elucidating example here is the ethanol production systems developed in the 1980s [55]. Key design parameters can be easily identified, for example, the geometry of the beads, diffusion in/out the beads, gas formation, water content, growth in the beads, material properties, infection, and sterilization (Table 9.6). Similar

TABLE 9.6 Key Design Parameters for Immobilized Cell Bioreactors

Immobilization Method	Key Design Parameter	Example
Alginate entrapped cells	Bead size	Entrapped yeast cells production of bioethanol
	Cell per bead particle	
	Bead porosity	
	Bead density versus media	
	Water/buffer content	
	Solvent/media resistance	
	Dissolution time	
	Fabrication cost	
	Toxicity	
	Sterilization possibility	
	Diffusion rate inside bead of gas and liquids	
	Durability for agitation and shear forces	
Microcarrier anchored cells	Bead size and cells per microcarrier particle	Surface-bound hybridoma cell culture producing monoclonal antibodies
	Porosity of microcarrier if porous	
	Surface charged groups	
	Diffusion rate to microcarrier	
	Durability of microcarrier	
	Time for loading the microcarrier with cells	

analysis can be applied on different microcarrier cultures such as hybridomas, surface growing CHO cells, and other cell types.

9.4.3 Bioreactors for Tissue and Stem Cell Cultures

The recent progress of tissue engineering has been accompanied with advanced technology for production of tissues in culturing devices. Also, for these applications the bioreactor concept is exploited. The main reason is, of course, that it is easy to reuse much of the experience of microbial and cell culture reactors for chemicals and biopharmaceutical production onto tissue cultivation. Aspects of oxygen transfer, mixing, containment, and barriers for infection as well as the engineering mathematical theory are applicable with little adaptation [2]. Tissue culture applications almost always presuppose a three-dimensional microstructure of the tissue. This is a striking difference from other bioreactors. Typical examples are cartilage cells, hepatocytes, and neural cells, while erythrocytes and other blood cells can be cultured in suspensions. When bioreactors are used for stem cells, additional requirements must be considered, such as phenotype stability and use of cocultures [59]. Conceptual design issues for tissue reactor devices have been discussed mainly from medical viewpoints but are still most relevant [1,60]. The static three-dimensional cultures in tissues generate partly other bioreactor concepts. For example, perfusion through membranes and hollow-fibers are useful configurations for tissue systems [4,61]. The perfusion is a transport process that is close to mass transfer of soluble nutrients, metabolites, and stimulating factors across other physical barriers than in a stirred suspension culture. The combination with scaffolds similar to immobilization structures as mentioned above [62] is a key issue in the tissue bioreactor design.

If the tissue cells are to be recovered, the procedures and cost and reuse aspects change the boundary conditions of the design alternatives.

Manufacture of stem cells is covered in a separate chapter of this book (Chapter 11). Since bioreactor engineering is indeed a central operation in stem cell processes, the technical design requirements we mentioned here will be further analyzed in the next chapter. Also, application of PAT and QbD methodology may play a key role in unifying the many aspects of modeling, analytics, and regulatory concerns [63]. Chapter 13 will highlight these aspects further.

Stem cell culturing is today *per se* a wide area covering very diverse cell types and culturing procedures [3,30]. Thus, bioreactors should be designed according to the special conditions of the purpose of the stem cell cultures. A variety of stem cell applications described in the literature span from hematopoietic stem cells in suspension cultures to advanced progenitor expansion systems in tissue-like bioreactor constructs [16].

The analytical instrumentation necessary for monitoring stem cell expansion and differentiation in the bioreactor must have the capacity to capture information about the histology, biomarker expression, and morphology of the cells [64,65].

9.4.4 Bioreactors for Plant Cell Cultures

Not yet so much used in industry but for many years in academic research pursued actively are bioreactions with plant cell cultures. The plant cell cultures are derived from callus preparations of the plant, then grown in suspensions or aggregates, and are able to express unique compounds, either from the secondary metabolism routes or, as has lately been done, by genetically engineered plant cultures [17]. The design of the plant cell bioreactors is complicated by the high viscosity of the culture, the accumulation of the product inside the cells, the supply of nutrients and oxygen, and the sterility requirements [9]. This has resulted in a number of elaborated bioreactors tailored for plant cell culture conditions [4]. This includes stirred tank bioreactor, pneumatic bioreactors, wave bioreactors, and also microbioreactors. By optimizing the mass transfer of media and oxygen, these designs have provided functional solutions to very vulnerable and in time extended plant cell cultures. Effects on oxygen consumption, aggregate formation, the difficult rheology of the plant cells, shear sensitivity, and foam and wall effects have been addressed. Using *in situ* sensing has been difficult due to limited optical transparency. Solutions with electronic nose and other noninvasive methods have become new design solutions [66].

For plant cell cultures, the ΣBioS are characterized by low growth rate, substrate uptake rates, and product formation rates. Critical issues are susceptibility to infections, optimal pH is favorable for many invasive bacteria, and plant cell cultures. The fouling effects on bioreactor walls and separation and sensor membranes are other interaction effects caused by plant cell cultures that restrict the choice of ΣTS alternatives. The anatomic view of the critical functional steps may be differently analyzed for plant cell bioreactors. The power consumption for agitation is high, which may lead to another consideration of the trade-off between cost and agitation equipment.

Recovery of the plant cell products such as intracellular secondary metabolites or recombinant proteins is more demanding. Although this is a design issue for the downstream process (see Chapters 8 and 15), there are aspects that can influence the bioreactor design (e.g., the use of media components is easier to separate).

9.5 CONCLUSIONS

The functional design analysis focuses on a number of critical issues and attributes for the selection of design alternatives. The outcome of the analysis may sometimes seem too obvious and brings about the sentiment that the conclusion of the analysis could have been much easier without the detailed disentanglement of already established facts.

This would be understandable, but still a wrong conclusion. The purpose of the detailed functional analysis is to uncover the nonestablished notions. These are found in those regions of the analysis that have not been devoted to intensive and challenging research as been the situation with genetic engineering, cell culture technology, and genomics.

Thus, the perhaps tedious analysis of how various technical means such as heat exchanger and impellers may possibly affect a sophisticated intracellular pathway analysis in the cell could be easily overlooked. We believe that several of the design matrices in this respect are self-explanatory. The outcome of the comparative analysis of the alternatives and the ranking of them will risk of becoming downprioritized at early R&D work.

The systematic cross-connection of the TrP stages and substages with the acting systems could in the same way cause the feeling of drawing obvious conclusion in a scholarly fashion, as a self-appreciating art of its own.

The breakdown of the systems into subsystems and smaller parts should be done by keeping the functions as much as possible and not changing the technical means too soon. This is sometimes difficult to realize 100%. When reaching the level of sub-subsystems, we recommend introducing the means for effectuating the functions.

REFERENCES

1. Martin, I., Smith, T., Wendt, D. (2009) Bioreactor-based roadmap for the translation of tissue engineering strategies into clinical products. *Trend. Biotechnol.* 27, 495–502.
2. Pörtner, R., Nagel-Heyer, S., Goepfert, C., Adamietz, P., Meenen, N.M. (2005) Bioreactor design for tissue engineering. *J. Biosci. Bioeng.* 100, 235–245.
3. King, J.A., Miller, W.M. (2007) Bioreactor development for stem cell expansion and controlled differentiation. *Curr. Opin. Chem. Biol.* 11, 394–398.

4. Eibl, R., Eibl, D. (2008) Design of bioreactors suitable for plant cell and tissue cultures. *Phytochem. Rev.* 7, 593–598.
5. Robinson, T., Nigam, P. (2003) Bioreactor design for protein enrichment of agricultural residues by solid state fermentation *Biochem. Eng. J.* 13, 197–203.
6. Atkinson, B., Manituva, F. (1991) *Biochemical Engineering and Biotechnology Handbook*, 2nd edition, Stockton Press, New York.
7. Nielsen, J., Villadsen, J. (1994) *Bioreaction Engineering Principles*, Plenum Press, New York.
8. Doran, P.M. (1995) *Bioprocess Engineering Principles*, Academic Press, London.
9. Shuler, M.L., Kargi, F. (2002) *Bioprocess Engineering: Basic Concepts*, 2nd edition, Prentice Hall, Englewood Cliffs, NJ.
10. Freshney, R.I. (1992) *Animal Cell Cultures: A Practical Approach*, IRL Press and Oxford University Press, Oxford.
11. Freshney R.I. (2005) *Culture of Animal Cells*, Wiley and Sons, Inc, New Jersey.
12. McNeil, B., Harvey, L.M. (1990) *Fermentation: A Practical Approach*, IRL Press and Oxford University Press, Oxford.
13. Funke, M., Buchenauer, A., Schnakenberg, U., Mokwa, W., Diederichs, S., Mertens, A., Müller, C., Kensy, F., Büchs, J. (2010) Microfluidic biolector, microfluidic bioprocess control in microtiter plates. *Biotechnol. Bioeng.* 107, 497–505.
14. Lee, P.J., Hung, P.J., Rao, V.M., Lee, L.P. (2006) Nanoliter scale microbioreactor array for quantitative cell biology. *Biotechnol. Bioeng.* 94, 5–14.
15. Singh, V. (1999) Disposable bioreactor for cell culture using wave-induced agitation. *Cytotechnol.* 30, 149–158.
16. Schmelzer E., Mutig K., Schrade P., Bachmann S., Gerlach J.C., Zeilinger K. (2009) Effect of human patient plasma *ex vivo* treatment on gene expression and progenitor cell activation of primary human liver cells in multi-compartment 3D perfusion bioreactors for extra-corporeal liver support. *Biotechnol. Bioeng.* 103, 817–827.
17. Huang, T.K., McDonald, K.A. (2009) Bioreactor engineering for recombinant protein production in plant cell suspension cultures. *Biochem. Eng. J.* 45, 168–184.
18. Revstedt, J., Fuchs, L., Trägårdh, C. (1998) Large eddy simulations of the turbulent flow in a stirred reactor *Chem. Eng. Sci.* 53, 4041–4053.
19. Simmons, M.J.H., Zhu, H., Bujalski, W., Hewitt, C.J., Nienow, A.W. (2007) Mixing in a model bioreactor using agitators with a high solidity ratio and deep blades. *Chem. Eng. Res. Design* 85, 551–559.
20. Rokem, J.S., Eliasson Lantz, A., Nielsen J. (2007) Systems biology of antibiotics production by microorganisms. *Nat. Prod. Rep.* 24, 1262–1287.
21. Tyo, K.E., Alper, H.S., Stephanopoulos, G.N. (2007) Expanding the metabolic engineering toolbox: more options to engineer cells. *Trend Biotechnol.* 25, 133.

22. Stephanopoulos, G.N., Aristidou, A.A., Nielsen, J. (1998) *Metabolic Engineering: Principles and Methodologies*, Academic Press, San Diego, USA.
23. Hutmacher, D.W., Singh, H. (2008) Computational fluid dynamics for improved bioreactor design and 3D culture. *Trend Biotechnol.* 26, 166–172.
24. Bracewell, D., Gernaey, K.V., Glassey, J., Hass, V.C., Heinzle, E., Mandenius, C. F., Olsson, I.M., Racher, A., Staby, A., Titchener-Hooker, N. (2010) Report and recommendation of a workshop on education and training for measurement, monitoring, modelling and control (M^3C) in biochemical engineering. *Biotechnol. J.* 5, 359–367.
25. Buchenauer, A., Hofmann, M.C., Funke, M., Büchs, J., Mokwa, W., Schnakenberg, U. (2009) Micro-bioreactors for fed-batch fermentations with integrated online monitoring and microfluidic devices. *Biosens. Bioelectron.* 24, 1411–1416.
26. Lye, G.J., Ayazi-Shamlou, P., Baganz, F., Dalby, P.A., Woodley, J.M. (2003) Accelerated design of bioconversion processes using automated microscale processing techniques. *Trend. Biotechnol.* 21, 29–37.
27. Lamping, S.R., Zhang, H., Allen, B., Ayazi Shamlou, P. (2003) Design of prototype miniature bioreactor for high throughput automated bioprocessing. *Chem. Eng. Sci.* 58, 747–758.
28. Ohyabu, Y., Kida, N., Kojima, H., Taguchi, T., Tanaka, J., Uemura, T. (2006) Cartilaginous tissue formation from bone marrow cells using rotating wall vessel (RWV) bioreactor. *Biotechnol. Bioeng.* 95, 1003–1008.
29. Duke, J., Daane, E., Arizpe, J., Montufar-Solis, D. (1996) Chondrogenesis in aggregates of embryonic limb cells grown in a rotating wall vessel. *Adv. Space Res.* 17, 289–293.
30. Jensen, J., Hyllner, J., Björquist, P. (2009) Human embryonic stem cell technologies and drug discovery *J. Cell. Physiol.* 219, 513–519.
31. Marks, D.M. (2003) Equipment design considerations for large scale cell culture. *Cytotechnology* 42, 21–33.
32. Ulrich, K.T., Eppinger, S.D. (2007) *Product Design and Development*. 3rd edition, McGraw-Hill, New York.
33. Mandenius, C.F., Björkman, M. (2010) Design principles for biotechnology product development. *Trend. Biotechnol.* 28, 230–236.
34. Derelöv, M., Detterfelt, J., Björkman, M., Mandenius, C.F. (2008) Engineering design methodology for bio-mechatronic products. *Biotechnol. Prog.* 24, 232–244.
35. Barrett, C.L., Kim, T.Y., Kim, H.U., Palsson, B.Ø., Lee, S.Y. (2006) Systems biology as a foundation for genome-scale synthetic biology. *Curr. Opin. Biotechnol.* 17, 488–492.
36. Heijnen, J.J., van Dijken, J.P. (1992) In search of a thermodynamic description of biomass yields for the chemotrophic growth of microorganisms. *Biotechnol. Bioeng.* 39, 833–858.

37. van Heijden, R.T.J.M., Heijnen, J.J., et al. (1994) Linear constraint relations in biochemical reaction systems I. Classification of the calculability and the balanceability of conversion rate. *Biotechnol. Bioeng.* 43, 3–10.
38. Mandenius, C.F., Brundin, A. (2008) Bioprocess optimization using design-of-experiments methodology (DoE). *Biotechnol. Prog.* 24, 1191–1203.
39. Haack, M.B., Olsson, L., Hansen, K., Eliasson, Lantz, A. (2006) Change in hyphal morphology of *Aspergillus oryzae* during fed-batch cultivation. *Appl. Microbiol. Biotechnol.* 70, 482–487.
40. Mosbach, K. (ed.) (1988) *Immobilized Enzymes and Cells: Methods in Enzymology*, Academic Press, New York., pp. 135–137.
41. Mandenius, C.F. (2004) Recent developments in the monitoring, modeling and control of biological production systems. *Bioproc. Biosyst. Eng.* 26, 347–351.
42. Warth, B., Rajkai, G., Mandenius, C.F. (2010) Evaluation of software sensors for on-line estimation of culture conditions in an *Escherichia coli* cultivation expressing a recombinant protein. *J. Biotechnol.* 147, 37–45.
43. United States Food and Drug Administration, (FDA). (2004) Guidance for Industry: PAT a Framework for Innovative Pharmaceutical Manufacturing and Quality Assurance. Available at http://www.fda.gov/cvm/guidance/published.html.
44. Mandenius, C.F., Graumann, K., Schultz, T.W., Premsteller, A., Olsson, I.M., Periot, E., Clemens, C., Welin, M. (2009) Quality-by-design (QbD) for biotechnology-related phamaceuticals. *Biotechnol. J.* 4, 600–609.
45. Schäpper, D., Alam, M.N., Szita, N., Eliasson, Lantz A., Gernaey, K.V. (2009) Application of microbioreactors in fermentation process development: a review. *Anal. Bioanal. Chem.* 395, 679–695.
46. Le Gac, S., van den Berg, A. (2010) Single cell as experimentation units in lab-on-a-chip devices. *Trend Biotechnol.* 28, 55–62.
47. Kumar, S., Wittmann, C., Heinzle, E. (2004) Minibioreactors. *Biotechnol. Lett.* 26, 1–10.
48. Kensy, F., Engelbrecht, C., Büchs, J. (2009) Scale-up from microtiter plate to laboratory fermenter: evaluation by on-line monitoring techniques of growth and protein expression in *Escherichia coli* and *Hansenula polymorpha* fermentations. *Microb. Cell Fact.* 8, 68, 1–15.
49. Zanzotto, A., Szita, N., Boccazzi, P., Lessard, P., Sinskey, A.J., Klavs, F. Jensen, K.F. (2004) Membrane-aerated microbioreactor for high-throughput bioprocessing. *Biotechnol. Bioeng.* 87, 243–254.
50. Ferstl, W., Klahn, T., Schweikert, W., Billeb, G., Schwarzer, M., Loebbecke, S. (2007) Inline analysis in microreaction technology: a suitable tool for process screening and optimization. *Chem. Eng. Technol.* 30, 370–378.
51. Sadani, Z., Wacogne, B., Pieralli, C., Roux, C., Gharbi, T. (2005) Microsystems and microfluidics devices for single cell oocyte transportation and trapping: towards the automation of *in vitro* fertilization. *Sens. Acuat. A Phys.* 121, 364–372.

52. Toh, Y.C., Lim, T.C., Tai, D., Xiao, G., van Noort, D., Yu, H. (2009) A microfluidic 3D hepatocyte chip for drug toxicity testing. *Lab Chip* 9, 2026–2035.
53. Qiu, Q.Q., Ducheyne, P., Ayyaswamy, P.S. (1999) Fabrication, characterization and evaluation of bioceramic hollow microspheres used as microcarriers for 3-D bone tissue formation in rotating bioreactors. *Biomaterials* 20, 989–1001.
54. Moo-Young, M. (ed.) (1988) *Bioreactor Immobilized Enzymes and Cells: Fundamentals and Applications*, Elsevier Applied Science, London.
55. Lother, A.M., Oetterer, M. (1995) Microbial cell immobilization applied to alcohol production: a review. *Rev. Microbiol.* 26, 151–159.
56. Goswami, J., Sinskey, A.J., Steller, H., Stephanopoulos, G.N., Wang, D.I. (1999) Apoptosis in batch cultures of Chinese hamster ovary cells. *Biotechnol. Bioeng.* 62, 632–640.
57. Hu, J.C., Athanasiou, K.A. (2005) Low-density cultures of bovine chondrocytes: effects of scaffold material and culture system. *Biomaterials* 26, 2001–2012.
58. Nettles, D.L., Elder, S.H., Gilbert, J.A. (2002) Potential use of chitosan as a cell scaffold material for cartilage tissue engineering. *Tissue Eng.* 8, 1009–1016.
59. Godara, P., McFarland, C.D., Nordon, R.E. (2008) Design of bioreactors for mesenchymal stem cell tissue engineering. *J. Chem. Technol. Biotechnol.* 83, 408–420.
60. Minuth, W.W., Strehl, R., Schumacher, K. (2004) Tissue factory: conceptual design of a modular system for the *in vitro* generation of functional tissues. *Tissue Eng.* 10, 285–294.
61. Simon, J. (2008) The status of membrane bioreactor technology. *Trend Biotechnol.* 26, 109–116.
62. Khademhosseini, A., Langer, R., Borenstein, J., Vacanti, J.P. (2006) Microscale technologies for tissue engineering and biology. *Proc. Nat. Acad. Sci. USA* 103, 2480–2487.
63. Mandenius, C.F., Björkman, M. (2009) Process analytical technology (PAT) and quality-by-design (QbD) aspects on stem cell manufacture. *Eur. Pharm. Rev.* 14, 32–37.
64. Kirouac, D.C., Zandstra, P.W. (2008) The systematic production of cells for cell therapies. *Stem Cell* 3, 369–381.
65. Thomas, R.J., Anderson, D., Chandra, A., Smith, N.M., Young, L.E., Williams, D., Denning, C. (2009) Automated, scalable culture of human embryonic stem cells in feeder-free conditions. *Biotechnol. Bioeng.* 102, 1636–1644.
66. Komaraiah, P., Navratil, M., Carlsson, M., Jeffers, P., Brodelius, M., Brodelius, P. E., Kieran, P.M., Mandenius, C.F. (2004) Growth behaviour in plant cell cultures based on emissions detected by a multisensor array. *Biotechnol. Prog.* 20, 1245–1250.

10

Chromatographic Protein Purification

This chapter discusses the design of chromatographic purification systems for proteins from the laboratory to the production scale. First, a background and the present state of the art of existing protein purification systems, including the basic chromatography methodology, are given (Section 10.1). Then, user needs of protein purification systems are analyzed and specified (Section 10.2). On the basis of these needs and specifications, biomechatronic design methodology is applied (Section 10.3) and exemplified for two cases, the design of (1) a large-scale system for manufacturing a specified protein (Section 10.4) and (2) a micropurification system using a lab-on-a-chip design (Section 10.5).

10.1 BACKGROUND OF CHROMATOGRAPHIC PROTEIN PURIFICATION

As an alternative to other unit operations for protein separation, such as membrane filtration or centrifugation, chromatographic separations are quite often the preferred choice of method [1].

The principle, to use a stationary phase that is able to selectively adsorb or partition the components in a mobile phase resulting in their separation, can be applied in a variety of ways [2,3]. Different stationary phases based on ligands

Biomechatronic Design in Biotechnology: A Methodology for Development of Biotechnological Products, First Edition. Carl-Fredrik Mandenius and Mats Björkman.

attached to a matrix where the ligand selectivity is based on ionic charged groups, hydrophobic substituents, or affinity molecules can be used. For partition separation, the matrix itself causes the selectivity.

Applications for chromatographic protein separation are widespread [3,4]. In laboratory-scale purification, the purpose is normally research driven; to purify a substance for subsequent identification and characterization or to investigate separation conditions to be applied later on a larger scale [1,5,6]. On a pilot or production scale, the purpose is to produce a protein of defined purity in an optimal and economical way [7,8].

The methodology of chromatographic protein separation dates back to the early 1960s and the history of chromatography long before that. Commercial systems appeared in the late 1960s. Thus, product design in this area has been pursued extensively for a long time and the significant experience and competence gained are today well established [6,7].

In Figure 10.1, four well recognized commercial systems are shown: the FPLC™ system from Pharmacia Biotech, the ÄKTA™ and Streamline™ systems from GE Healthcare, and pilot-scale systems from BioRad. As evident, the physical design solutions in these products are similar. The basic chromatographic principles remain unchanged in these setups although continuously improved and refined. This possibly means that the ideal design has been reached today. Or is there still room for more significant improvements or new conceptual design solutions?

A standard protein purification setup for large-scale production is shown in Figure 10.2. The chromatographic column is fed with either a feed of raw materials containing the proteins to be separated or a buffer for elution of target product(s) and/or impurities, a regeneration buffer or an equilibrium buffer.

Along the flow line, several sensors or gauges monitor the fed media: two air sensors check the content of oxygen, a pressure sensor controls the hydrodynamic pressure in the flow, a flow meter gauge adjusts the flow rate of the pump, after the column pH, conductivity, and UV-absorption are monitored. Air bubbles are removed in a trap device and particles are removed by a microfilter before the column. The product protein, residual proteins, and waste are collected in tanks by switching valves from predetermined sequences. This is done with the help of sensor signals or manually. Not shown in the figure are the automation circuits controlling the system from set procedures in a control software.

In a laboratory-scale setup with a smaller column, miniaturization of equipment parts would be necessary. Also, the product tanks would be replaced by a fraction collector with small vials or titer well plates.

The stationary phase chosen for the column is a decisive component of the system. The choices are typically between a size exclusion gel, an ion exchange resin, an affinity hydrogel, a gel with hydrophobic ligands, or

Product	System	Vendor
	ÄKTA™ Explorer. A system introduced in late 1990s as a replacement for FPLC. The system design is more compact allowing scale scale to be applied. This makes the system very useful for scaling up of a chromatographic unit operation.	GE Healthcare (USA/S)
 BioRad Easy Pack Column BioRad GelTec Column BioRad InPlace Column	Pilot- and production scale. Similar system approach for pilot- and production scale chromatography. Design comparable.	BioRad (USA)
	Streamline™ Expanded bed-based system allowing moderately particulate media to be processed.	GE Healthcare (USA)
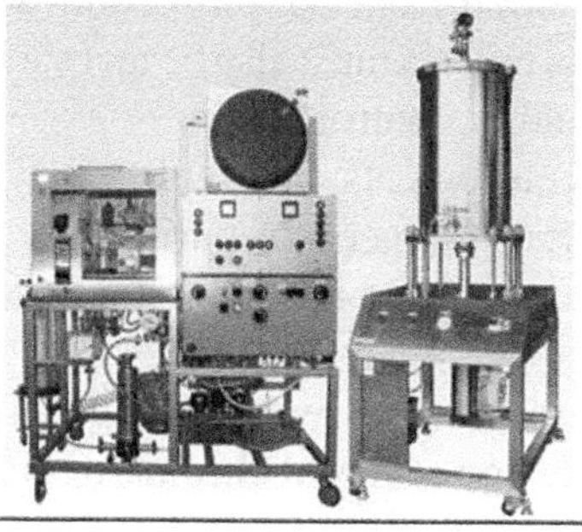	A complete system with chromatography column and control units.	KronLab Chromatography (D)

Figure 10.1 *Four commercial protein purification systems (ÄKTA purifier, Streamline, and BioRad and KronLab pilot systems). Reproduced with permissions from GE Healthcare, BioRad, and KronLab Chromatography.*

a reversed-phase matrix. The varying capacity for adsorption of these stationary phases in terms of loading, separation ratio, resolution, and so on results in quite different requirements for flows and column sizes. These are

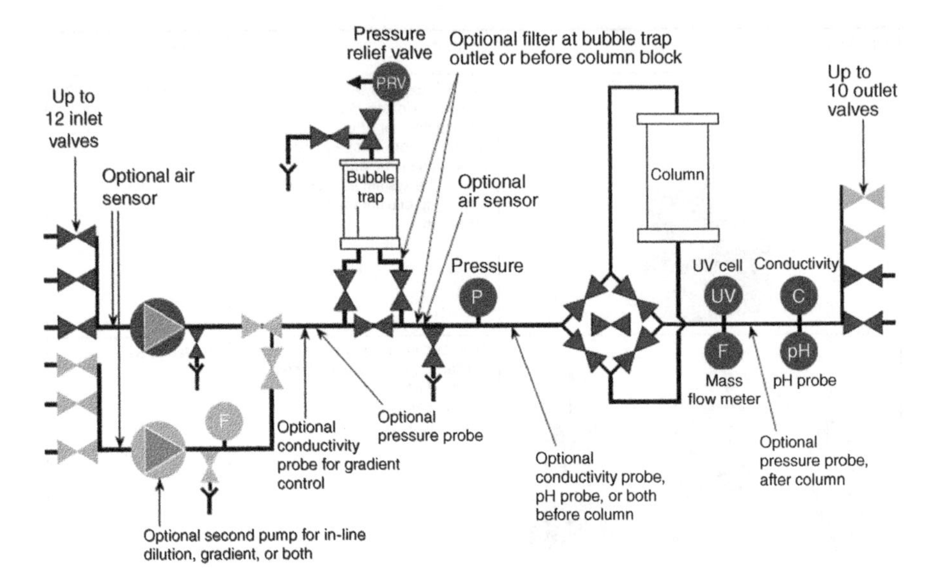

Figure 10.2 *Basic setup of a large-scale purification system (automation not shown). Reproduced with permission from BioRad.*

important design constraints that must be extensively evaluated for each protein separation methods to be developed.

When using the chromatographic system as a unit operation of a bioprocess, thus being a step in the process dependent on proceeding unit operation, for example, a centrifuge, a membrane filter, or another chromatographic column, the ability to cope with the raw material variation from the proceeding step becomes a key issue. A number of other process criteria must also be met such a total production economy of the plant and quality control issues.

In the laboratory-scale system, these requirements are replaced with adaptability and precision of the system. Often the need of purifying very minute amounts of a protein may become crucial for the laboratory-scale application.

The scientific literature on protein purification is by now huge. The number of scientific articles with studies on design issues of chromatographic systems is very significant and not possible to cite or discuss in detail in this chapter. A few key references are given (in which further cross-references can be found) where essential facts about critical design criteria are accounted for [9–14].

In several published articles, it is clear that a systematic design methodology is developed and used (e.g., [13,15]). However, in none of these a mechatronic-inspired methodology is applied although sometimes the approaches are quite close and result in useful creative designs.

10.2 SPECIFICATION OF NEEDS FOR PROTEIN PURIFICATION SYSTEMS

From the account above, it may appear that needs for protein purification systems are too dispersed for setting up a general specification. However, also at this stage of designing it is possible and useful to provide guidelines for structuring the analysis of needs. Since the theory of chromatography is thoroughly treated in terms of engineering parameters, these can be easily listed and ranges for their target values established [16–20, 21]. Desirable levels of loading capacity of the column, throughput, separation factors, plate number of column, volume of column including width and length, and flow rates can be defined for the separation method [22–26]. Several of these parameters are interdependent and can be adapted from the basic textbooks on chromatography [27–29]. Thus, optimization of the method comes as a development stage in the setup that the design should anticipate.

The needs imposed on the separation gel itself should also be included in the specification table. For affinity gels, ranges of affinity parameters (e.g., affinity constants K_a and K_d and the affinity rate constants k_{on} and k_{off}) can be defined. The loading capacity of the gel (e.g., gram protein per gram coupled gel) can be defined, for example, using literature values.

For hydrophobic interaction chromatography (HIC) gels and ion exchange resins similar gel capacity parameter could be given.

More difficult parameters to define are rheological properties, robustness to variation of the raw materials, nonspecific adsorption effects, and longevity of the column and the gel. These parameters are highly dependent on the variation of the materials and must necessarily be tested in experiments.

Needs that are less quantifiable such as convenience of operation, durability of equipment, which online sensors/meters should be included, which control loops should be established, what level of automation should be integrated are set by yes/no or with rough estimates (high/medium/low).

For example, sterility barriers are in pharmaceutical production regulated by authorities and are therefore obviously necessary design criteria. Established sterility tests can be used in the specification. GMP rules according to existing guidelines can be included.

Construction materials should be specified thereby counteracting potential effects from aggressive raw materials and elution, regeneration, and equilibration buffers. These media may have detrimental effects on materials used for columns, valves, and pumps. The media composition is to large extent known (acidic and alkaline solutions and solvents). Thus, stainless steel columns may be necessary and Teflon valves and tubing required to satisfy robustness and longevity needs.

TABLE 10.1 List of Needs and Target Specification for Protein Purification Systems

Needs → Metrics	Target Value: Laboratory-Scale System	Target Value: Production-Scale System	Units
Performance			
High throughput	1	2000	g protein/h
High purity	99	97	Percentage purity
High mass load	0.1	0.1	Protein/g gel
Good recovery	98	98	Percentage recovery
Reasonable cost	Not applicable	30	Percent cost of manufacturing
Good resolution	95	95	Percentage
High volume load	10	10	g/L gel
High flow rate	>20	>20	Columns per day
Geometry/column size			
Aspect ratio	10:2	10:2	H/W ratio
Column size	0.01	100	Column volume (L)
Monitoring and control			
pH controlled	Yes	Yes	Yes/no
Conductivity controlled	Yes	Yes	Yes/no
Temperature controlled	Yes	Yes	Yes/no
Pressure controlled	Yes	Yes	Yes/no
Air level in media controlled	Yes	Yes	Yes/no
Flow rate controlled	Yes	Yes	Yes/no
Protein monitored at elution	Yes	Yes	Yes/no
High hygiene standard			
Contained plant	No	Yes	Yes/no
Personnel protected	No	Yes	Yes/no
Particle count in atmosphere			Particles/m^3
Media pretreatment			
Air bubbles removed	Yes	Yes	Yes/no
Particle removal	Yes	Yes	Yes/no
Sanitation facilities			
Chemical sanitation	Yes	Yes	Yes/no
UV sanitation	No	Yes	Yes/no
Steam sanitation	No	Yes	Yes/no
Automated operation			
Complete computer control	No	Yes	Yes/no
Semicontrol	Yes	No	Yes/no
Equipment durable			
System compact	Yes	No	Max. volume of unit
Equipment in stainless steel	No	Yes	Yes/no
Equipment in glass/plastics	Yes	No	Yes/no

Table 10.1 lists a collection of needs and specifications of a laboratory-scale and a production-scale system in a List of Needs and Target Specification.

Note that it is not only a difference in scale that distinguishes the two systems but also their different purpose and how they are used in their application.

Certainly, the list of needs could be easily extended further. Once the actual design case is chosen, some needs may be removed and other added. The metrics given here for the target specification are examples of possible values, most of them taken from various literature reports or the authors' own experiences. In several of the metrics, real experiments are necessary to carry out.

It should also be noted that for some specifications no differences exist for laboratory-scale and production-scale targets. For example, several of the monitoring sensors are independent of scale and application. For other specifications, for example, GMP requirements and hygiene needs, it may be quite different.

For a production-scale system, it is normally only one type of gel and one product that are dealt with. For the laboratory-scale system, different gels/columns must be possible to use in the system. Also, a range of different raw materials should be possible to test and be used on small scale.

The need for automation may be unnecessary for small systems, but must most probably be mandatory for large-scale operation applications.

10.3 DESIGN OF PURIFICATION SYSTEMS

10.3.1 Generation of Design Alternatives

The content of the List of Needs and Target Specification can generate quite a number of design solutions including several nonchromatographic alternatives. The Concept Generation Chart of Figure 10.3 depicts 10 design suggestions based on Table 10.1 target specifications.

First, a highly schematized structure valid for all generated alternatives below is shown in a Basic Concept Component Chart. It is based on a target protein and the impurities as two entities, an elutor function (usually an eluent molecule), a separator function, a collector for pooling or sampling fractions with separated biomolecules, and a detector function for identifying targets and impurities. The subsequent alternatives are all variations and extensions of this motive in the Permutation Chart.

Alternatives A, B, and C are examples of design solutions that utilize batch procedures. Batch processing is not chromatographic and thus not under the title of the chapter although they can be easily squeezed-in into the criteria on the specification. These are still shown here to bring to the readers' attention that they should always consider alternatives widely and be prepared to cross boundary restrictions if suitable.

The same can be said about alternatives D, E, and F that are all using membrane operations to separate the target protein. In a way, the A–F alternatives come in some respects close to chromatography design solutions.

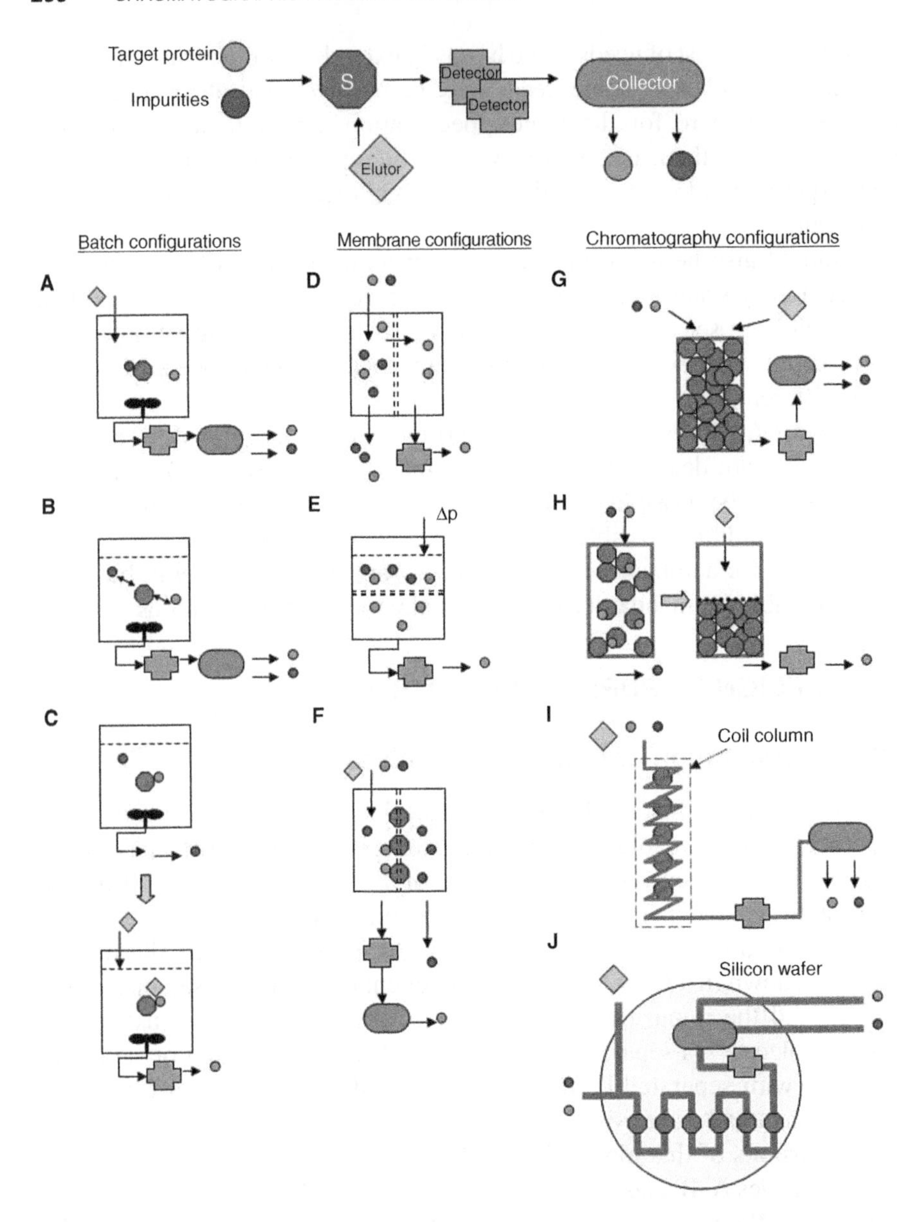

Figure 10.3 *A Concept Generation Chart for the protein purification system. The top shows the Basic Concept Component Chart and below is a Permutation Chart for 10 design concepts generated from specification. (S) is the design function that causes separation. Note that only the concepts G–J are chromatographic and fit the chapter title. It is concepts G and J that are described in the case examples below.*

Alternative A is a batch reactor where a separator function is immersed into the medium together with the target protein and impurities. The separator prefers the target protein while the impurities pass by. The detector checks the performance and assists the collector to pool the molecules.

Alternative B reverses the preference of the separator, that is, takes up the impurities and allows the target protein to pass by.

In alternative C, the batch procedure is carried out stepwise. First, the molecules are adsorbed to the separator. In the second step, the molecules are selectively separated by additions of eluent(s).

In alternative D, a selective membrane is introduced that can sieve the target from the impurities. It is set up in a tubular flow column.

Alternative E is the design solution but here a pressure drives the transport across the selective membrane.

In alternative F, the membrane is reinforced with a separator (e.g., an antibody ligand) that supports the separation at the membrane surface.

The chromatographic alternatives have all a mobile phase containing the target protein and impurities and a stationary phase with the separator.

Alternative G is a typical chromatography column and alternative H is an expanded bed affinity chromatography system framed into this conceptual view.

In alternative I, the column is a capillary coil filled with a separator function, either grafted to the column wall or filled in the void of the column.

Alternative J is an attempt to try to realize the system on a silicon wafer with lab-on-a-chip technology. It is consequently a significant miniaturization of the system that leads to interesting implications.

Additional concept alternatives are no doubt possible. We think that these 10 should bring out the scope of the methodology.

In the following, we will pursue the chromatography inspired design approach in its entirety.

10.3.2 Screening the Design Alternatives

As in previous chapters of this book, we compare the generated alternatives in a Concept Screening Matrix (Table 10.2). We have here limited the screening to those alternatives that involve chromatographic purification and added two other alternatives (A and D) for comparisons. As selection criteria, the needs from the List of Needs and Targets Specifications are used. Some of the selection criteria are difficult to compare. Then "0" scores are given (e.g., the criteria are excluded).

The screening outcome ranks alternatives G and I highest. Interestingly, the expanded bed alternative is ranked fourth. And the batch and membrane alternatives are both inferior to the chromatographic solutions.

TABLE 10.2 Concept Screening Matrix for Selection Criteria (Correct Criteria To Be Checked)

	Concepts					
Selection Criteria	Concept A	Concept D	Concept G	Concept H	Concept I	Concept J
Performance						
High throughput	−	+	+	0	+	+
High purity	+	+	+	+	+	+
High mass load	0	0	+	+	+	+
Good recovery						
Reasonable cost	0	0	+	+	+	+
Good resolution	−	−	+	+	+	+
High volume load	−	−	+	+	+	+
High flow rate	−	+	+	0	+	+
Geometry/column size						
Aspect ratio	0	0	0	0	0	0
Column size	−	0	+	+	+	+
Monitoring and control						
pH controlled	+	+	+	+	+	+
Conductivity controlled	+	+	+	+	+	+
Temperature controlled	+	+	+	+	+	+
Pressure controlled	0	0	+	0	+	0
Air level in media controlled	0	0	0	0	0	0
Flow rate controlled	+	+	+	+	+	+
Protein monitored at elution	+	+	+	+	+	+
High hygiene standard						
Contained plant	−	−	+	0	+	+
Personnel protected	0	0	0	0	0	0
Particle count in atmosphere	0	0	0	0	0	0
Media pretreatment						
Air bubbles removed	−	−	+	+	+	+
Particle removal	+	+	+	+	+	+
Sanitation facilities						
Chemical sanitation	+	+	+	+	+	+
UV sanitation	−	+	+	+	+	+
Sum +'s	8	11	19	16	19	18
Sum 0's	7	8	4	8	4	5
Sum −'s	8	4	4	4	0	0
Net score	0	7	19	16	19	18
Rank	6	5	1	4	1	3

10.3.3 Analysis of the Generated Alternatives for a Chromatography System

All design alternatives that are based on chromatography will fit into the Hubka–Eder map in Figure 10.4. The map may appear as a depiction of the process scheme of Figure 10.2, but it is worth noting that other quite different alternative designs are also possible to enclose in the Hubka–Eder mapping.

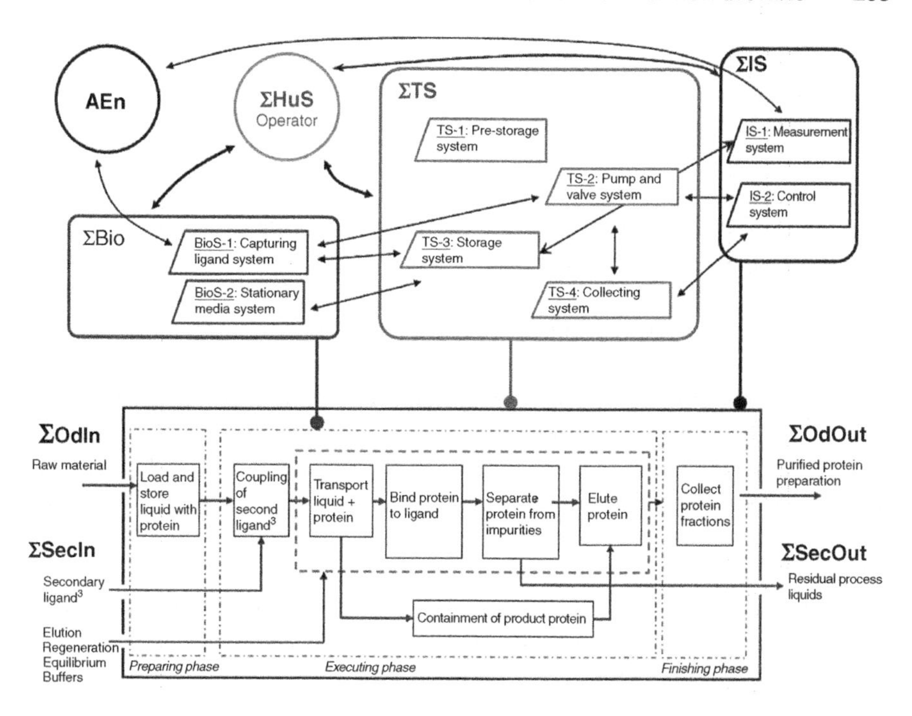

Figure 10.4 *Hubka–Eder map of a protein purification system. Due to space limitation, the ΣM&GS is not shown in the diagram.*

Let us again remind about that when following the Hubka–Eder mapping approach based on operands, transformations, and interactive supportive systems, including the biological systems, a generic and comprehensive model should be defined that merges special and constrained methods and procedures from different knowledge and competence backgrounds, such as electrical circuit diagrams, mechanical CAD drawings, or bioprocess flow-sheeting [1,6]. Then, this makes it easy to obtain an overview of the interactions between the biological and the technical functions, which in turn provides a basis for design and evaluation.

The Hubka–Eder map in Figure 10.4 shows the key chromatographic operations are encased within the transformation process boundary onto which the technical (ΣTS), biological (ΣBioS), information (ΣIS), and human (ΣHuS) systems, as well as the active environment (AEnv), interact.

Here, the ΣTS include all functions such as pumps, valves, sensors, and control systems that will be contained in the final designed product, the ΣBioS with chromatographic gels and ligand biomolecules or organic molecules for these gels that will carry out affinity and hydrophobic interaction chromatography with the target molecules to be separated. The ΣHuS include, for example, the technician that operates the system; the ΣIS include the methods and protocols being set up (today replaceable by computer expert software programs); and the AEnv includes external and unanticipated effects on the product (i.e., the transformation system), such as temperature, pH, and pressure.

The most relevant transformation properties appear at the next level of detail, where the favorable values of the ligand–ligate affinity rates are attained and where high separation factors and efficient plate numbers of columns are reached. What is of utmost interest at this level of detail is how the properties and transformation process by the biological system, as defined here, interact with the other subsystems of the technical system; that is, the active environment, the information systems, and other technical subsystems such as pumps, valves, and detectors.

It could be noted that the transformation process (TrP) is provided with the option regenerating of a specific ligand, such as a monoclonal antibody, by binding to a capture gel, for example, a Protein G gel. This is applicable only for affinity chromatography applications.

The management and goal systems (ΣM&GS) as used in other chapters is here excluded in order to simplify the description and since there are few interesting design issues to bring up around it.

The map in Figure 10.4 is the result of a detailed analysis of the functions and operational procedures of the protein purification system. Although seeming obvious at first, it reveals the functions and consecutive operations of the technical system in a different and clearer way than would be possible with a conventional design drawing, thereby illustrating the advantage of Hubka–Eder reasoning.

We are especially interested in a better understanding of the role of the biological molecules in the design. Therefore, we include a map that zooms-in at the ΣBioS and ΣTS systems (Figure 10.5).

One of the sub-subsystems of the BioS-1, in the map referred to as BioS-1.1, is the release of the target protein from the ligand (Figure 10.6). This is *per se* a function of the ligand subsystem. Thus, the sub-subsystem is not necessarily an object but a characteristic of the ligand. This is important since we want to understand how other subsystems of the design may influence this function. For example, it may be relevant to ask if certain properties of the stationary phase or the pump and valve subsystem affect the release.

The second sub-subsystem (BioS-1.2) in the map is the function of the ligand to allow efficient coupling to the stationary phase. This includes the chemical structure of the ligand and the matrix of the stationary phase. This leads us to ask if other subsystems could also influence the design?

The third sub-subsystem (BioS-1.3) is the function of the ligand to withstand all operation conditions executed. This certainly involves the technical subsystems and how these could impose influence on the ligand/matrix.

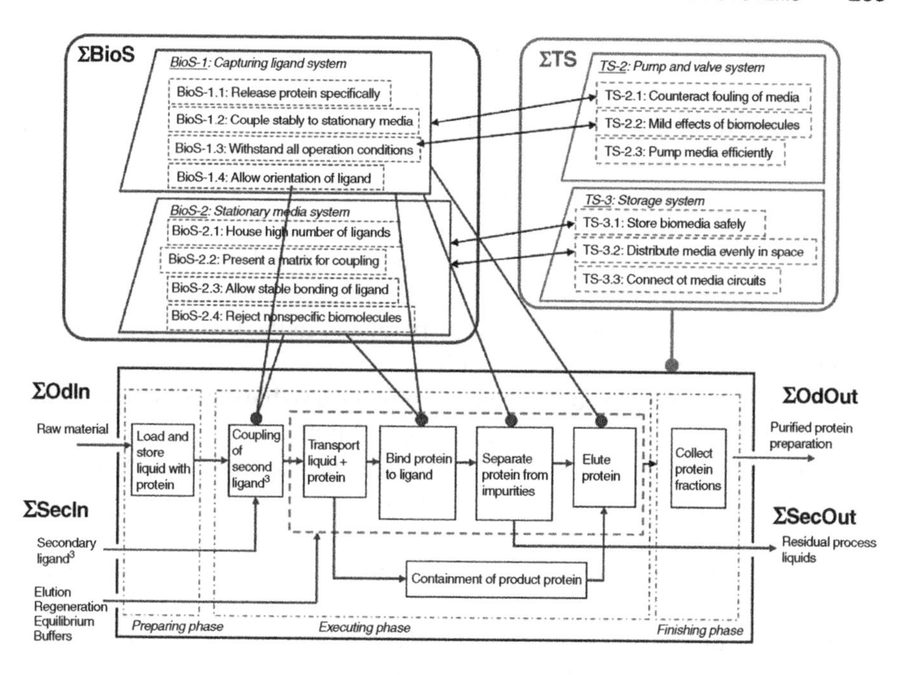

Figure 10.5 *Hubka–Eder map of a protein purification system highlighting essential parts of the biological technical systems and their interactions. Adapted from Ref. [2].*

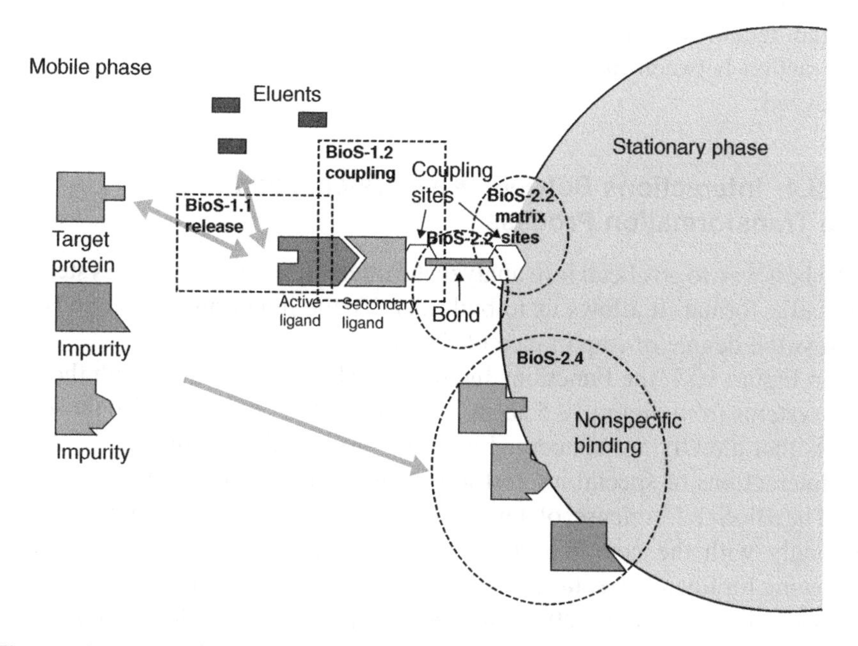

Figure 10.6 *The events occurring at the interface between the mobile phase and the stationary phase in a chromatographic column.*

The fourth sub-subsystem (BioS-1.4) is the functionality of the ligand to expose its interactive site to the target protein. This might involve orientation of the ligand toward the mobile phase.

The stationary phase subsystem BioS-2 has also four sub-subsystems that are functional characteristics of the phase.

BioS-2.1 concerns the capacity of the phase to house as many ligands as possible in order to achieve a high efficiency of the gel per volume.

BioS-2.2 concerns the presentation of binding sites for the coupling of the ligand. Obviously, this function is directly related to the BioS-1.2 above and depends on the type of ligand available.

BioS-2.3 is the function of the phase to provide a stable environment for the ligand. Again, this strongly depends on ligand type.

Finally, BioS-2.4 is the function of the phase to negate nonspecific binding of components in the mobile phase, predominantly the raw materials. However, nonspecific binding properties should, despite the mobile phase component interactions, also look into what other sub-subsystems may do, or may not do, to improve the functionality.

Only parts of the subsystems in the ΣTS are here showed and the ΣBioS could certainly be extended further with additional subsystems.

Here, the functionality of the ligands and the chromatographic stationary phase is highlighted. We wish to know more details about how different configurations of the design influence different design solutions. Would certain technical subsystems be more critical than others? Are there cross-interaction between the biological subsystems? This requires to be further analyzed.

10.3.4 Interactions Between Key Systems and the Transformation Process

The best way to go ahead is to analyze the interactions between the systems in a matrix format. It allows us to both identify the important interactions and assess the degree of importance of them.

In Figure 10.7, the Functions Interaction Matrix is lined out with the sub-subsystems from Figure 10.5 above. The matrix has on its axes also those TrP steps that directly are concerned with the biological systems.

Interactions of special interest to comment on are the following.

The BioS-1.1 (release of target protein from the ligand) interacts very strongly with the coupling stability, withstanding operational conditions, allowing for ligand orientation and stable bonding to the matrix gel according to the indicator in the matrix (highest level, 5). This should be understood so that the protein release will be very much affected by these properties of the ligand itself gel. Thus, special efforts should be devoted to these properties

Main Systems	Subsystems	ΣBioS								ΣTS			TrP						
		Release protein specifically	Couple stably station. media	Withstand oper conditions	Allow orienta-tion of ligand	House high num. of ligands	Present matrix for coupling	Allow stable bond. of ligand	Reject nonspec biomolecules	Counteract foul. of media	Mild effects of biomolecules	Distribute evenly in space	Load/Storage of raw material	Transport of liquid + protein	Bind protein to ligand	Separate prot. fr. impurities	Elute protein	Containment of product protein	Collect protein fractions
Σ BioS	Release protein specifically		5	4	4	1	3	3	5	4	4	3	1	1	5	5	5	1	1
	Couple stably to stationary media	4		5	3	4	4	5	2	3	1	5	4	2	5	5	4	5	4
	Withstand all oper. conditions	3	5		4	4	4	5	5	4	4	3	4	2	4	5	4	3	3
	Allow orientation of ligand	5	5	5		5	5	2	2	3	1	4	5	1	3	5	5	4	2
	House high number of ligands	4	2	2	4		5	3	5	5	3	4	5	1	5	5	4	4	3
	Present matrix for coupling	5	5	5	5	5		5	5	3	1	2	2	3	1	4	5	5	4
	Allow stable bonding of ligand	5	5	5	5	5	5		5	4	1	4	3	3	1	2	3	3	3
	Reject non-specific biomolecules	3	4	4	4	4	5	2		3	1	3	3	3	1	2	4	4	3
Σ TS	Counteract fouling of media	5	4	4	1	5	3	3	4		5	1	4	4	1	2	4	4	1
	Mild effects of biomolecules	3	5	1	1	2	2	4	2	5		1	1	5	5	5	5	5	4
	Distribute evenly in space	1	1	1	1	1	1	1	1	4	2		5	5	5	3	4	4	3
TrP	Load/Storage of raw material	1	1	1	1	1	1	1	1	5	5	5		5	3	2	4	4	3
	Transport of liquid + protein	4	4	4	4	4	4	4	4	4	4	4	4		4	4	5	5	4
	Bind protein to ligand	1	1	1	1	1	1	1	1	1	5	1	1	4		4	5	5	4
	Separate prot. fr. impurities	1	1	1	1	1	1	1	1	1	1	1	1	5	5		5	5	5
	Elute protein	4	4	2	2	5	4	4	1	2	1	5	3	5	3	4		5	3
	Containment of product protein	1	1	1	1	1	3	3	3	3	1	5	4	5	5	4	5		4
	Collect protein fractions	5	5	5	5	5	2	2	4	2	2	2	2	5	5	5	5	5	

Figure 10.7 *Functions Interaction Matrix for a protein purification system. The numbers index the strength of the interaction (5 = very strong; 4 = strong; 3 = medium; 2 = weak; 1 = very weak or nonexisting).*

in the design. The matrix also shows strong but not very strong interactions with the TS subsystems and the TrP steps.

The BioS-1.2 is devoted to the function of stably immobilizing the ligand to the stationary media. This chemical design matter has been a key issue in chromatography and many methods and problems with a variety of immobilization technique are possible to consider [9].

We do not go through all of the 18 × 18 interactions in the matrix (it may be an exercise for the interested reader to critically analyze the ranking of the interactions).

Note that squares indicated with number 1 are considered insignificant or nonexisting. This is something that should be cautiously done and repeated – not to delete any possible interaction before it is really made certain that it do not exist. Oblivion of possibilities may have serious consequences for the design functions.

In the assessment done in Figure 8.7 as many as 91 of 324 possible interactions are ranked as very strong (5) and 72 as strong (4). This should direct the design work and help the designers to make the work more focused and time effective.

Here, we have brought up only the biomolecular aspects. The matrix analysis should, however, be made with all systems, which we recommend for a complete design task.

10.4 UNIT OPERATION PURIFICATION IN A FVIII PRODUCTION PROCESS (CASE 1)

The case illuminates the design of an immunoaffinity chromatographic step in a downstream process for production of the blood coagulation protein Factor VIII. The recombinant human protein is produced in a human embryonic kidney (HEK) cell culture using a $1\,m^3$ bioreactor and the affinity chromatography operation is proceeded by a continuous centrifugation step. The purpose of the immunoaffinity step is to remove inappropriate glycosylation forms of FVIII formed in the HEK cell culture.

The conceptual design mission is to construct the affinity chromatography step from scratch. The target specification in Table 10.1 is applicable with a few changes. The range of impurity variation can be more accurately defined. The purity that must be accomplished in the affinity step can be set based on the specification of the subsequent formulation step.

The monoclonal antibody used in the immunoaffinity gel has distinct specificity to the desired FVIII glycoform. Thus, choice of method seems quite appropriate provided the functions of the ΣBioS shown in Figure 10.5 are possible to realize.

The charging and recharging of monoclonal antibody specific for FVIII on a Protein G gel (Sepharose G) is included in the TrP as shown in the map in Figure 10.8.

Effects of nonspecific adsorption need to be thoroughly investigated in experiments. Alternative gels should be considered. The elution buffer and other media coming in contact with the stationary phase and their reactions with it under the conditions created by the ΣTS should be taken into account.

The elution could be a pH gradient. Useful information about the eluted FVIII peak for pooling the protein would require a monitor. Due to lack of reliable specific FVIII monitors (e.g., an immunosensor), a conventional UV-monitor at 280 nm is used. This probably leads to some inefficiency and loss of FVIII when pooling the fractions from the column outlet. Thus, the interaction matrix should here have a high number to indicate the importance of this interaction (not shown).

The stability not only of the bonding of Protein G but also of the monoclonal IgG to Protein G under the conditions is already high in the functions interaction matrix above. This is confirmed in this case example.

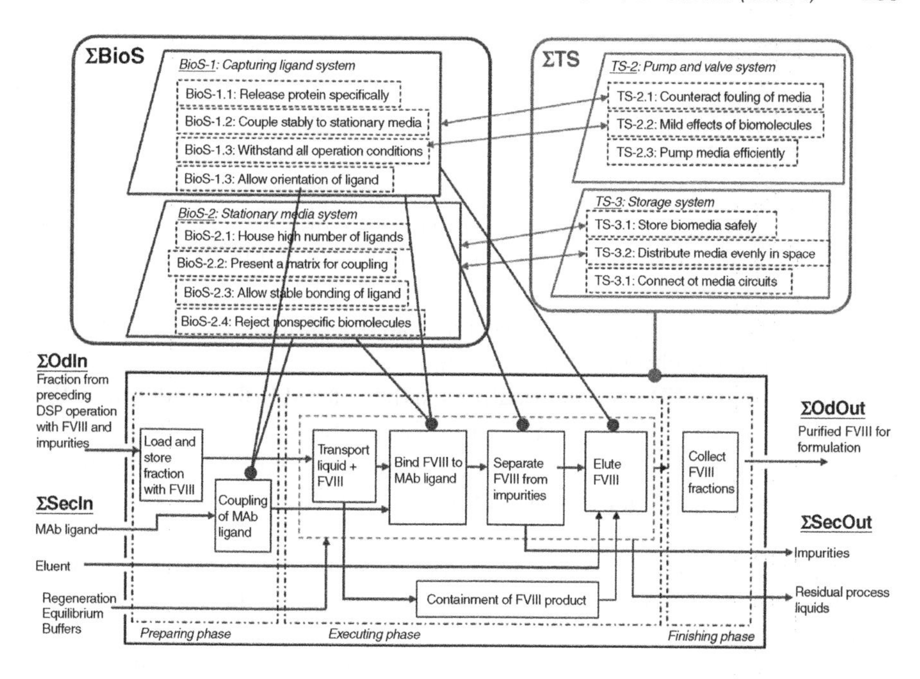

Figure 10.8 *Hubka–Eder map for the case 1 example where FVIII is purified in an immunoaffinity chromatography operation.*

The nonspecific adsorption of impurities (including the adverse glycoforms of FVIII) will reduce the throughput in the step. All the reactions shown in Figure 10.6 have strong or very strong relevance for this. An upgrading of interaction importance is justified in Functions Interaction Matrix.

The long-term stability of the ligand (due to degradation, slow desorption, and so on) determines the need for recharging the stationary phase. This means that the TrP step in the preparing phase is carried out only intermittently. Generation of information of this need of recharging would benefit from a subsystem function in the ΣIS. This could, for example, be a sensor of software type where the peak performance could assess right time of recharge.

Thus, some modifications have been necessary to be done in the Hubka–Eder map in Figure 10.8 but only to a limited extent.

10.5 MICROPURIFICATION SYSTEM BASED ON A MULTICHIP DEVICE (CASE 2)

This case describes the design of a multipurpose miniaturized protein purification device based on chromatographic separation. It is an attempt to realize the design alternative J in Figure 10.3. The design is based on the possibility to use silicon wafers or organic polymers with integrated structures.

The needs and target specifications would follow column 3 in Table 10.1 for a flexible laboratory-scale system.

As seen in the figure, a channel in the wafer constitutes the chromatographic column. This is filled with stationary phase carrying the separation ligand, or the ligand is grafted to the wall of the channel. A sensor, for example, a fiber optical UV-LED sensor, is integrated into the chip after the channel. Directly after the channel, a fraction collector is merged, for example, using an array of pneumatic microvalves.

The structured chip could be mass fabricated separately at a very low cost and shipped to the chromatography user.

Considering the ΣBioS functionalities in the general Hubka–Eder map, the coupling of ligands would benefit from being performed in the designed device. Therefore, the Hubka–Eder map is modified with a preparing phase where immobilization of the users' choice of ligand is done with the supplied chips.

In order to make the system adaptable to scale, chips should be run in parallel with up to 100 chips. This requires a disperser unit that requires special attention to design.

The modified Hubka–Eder map in Figure 10.9 shows several of the design changes in the TrP phases. The input and output operands are here rather

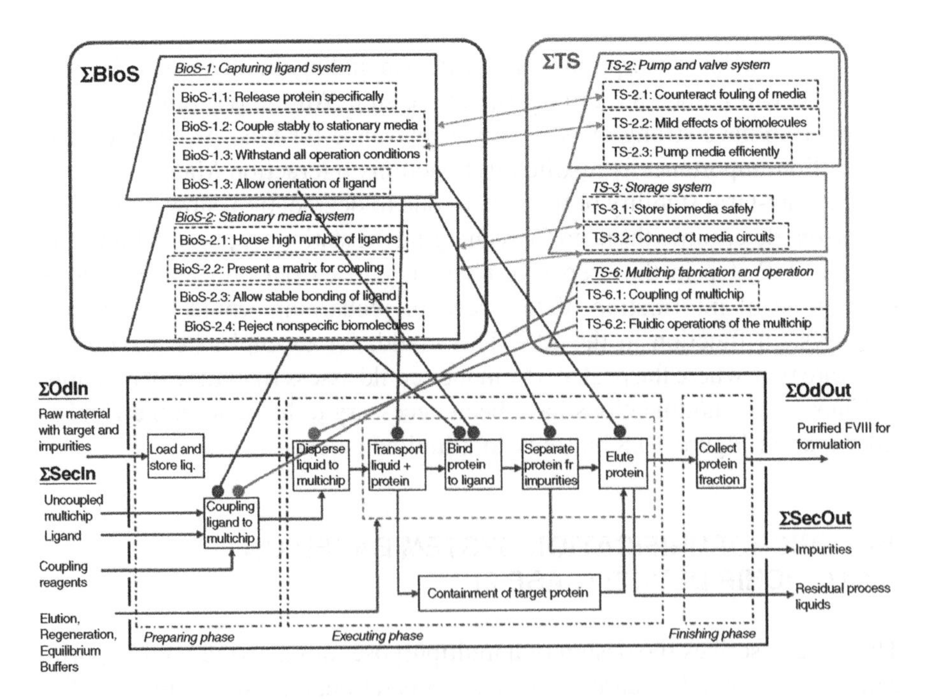

Figure 10.9 *A Hubka–Eder map for a multimicrochip purification system.*

different. The ΣTS is provided with a subsystem function for the chip preparation and TS1 with a subsystem for operating the dispersion subprocess.

A powerful option with the chip is to integrate sensors. Today's state of the art of the lab-on-a-chip technology furnishes a multitude of sensor chip applications where different electronic and optical sensors are integrated in-line (e.g., pH, pO_2, conductivity, NIR, UV, fluorescence, visible absorption). These possibilities are attractive for a chromatographic separation microchip. They should form subsystems of the ΣTS entity and be analyzed cross-connectively with the other systems. The miniaturization results in much shorter optical path lengths. Thus, the sensitivity must cope with this to be useful.

The capacity of each chip is minute compared to a conventional chromatography column. This is a function of the volume of the chip channel and the amount of gel that can be filled in it. The rheology of the flow is typically laminar and the space velocity should follow what is characteristic for large-scale columns.

The interactions in Figure 10.9 also highlight nonspecific effects. The column channel in silicone (or organic polymer) presents another interface. The volume/surface area ratio is quite different making the influence of the channel surface much more significant than what is experienced in a conventional column. Counteractions must be considered, such as coating the channel surface with suitable surface-active substituents. This may be specified in the chip fabrication method; otherwise, it can be included in the preparing phase as an additional subprocess in the TrP.

10.6 CONCLUSIONS

Design of large-scale protein chromatography purification is already well established. It may therefore seem unnecessary to try to create additional design guidelines for further development.

Nevertheless, conceptual mechatronic design theory is not introduced in this area of bioengineering and should in our view be considered.

In retrospect, many of the separation methods amply described in the past might have been realized earlier than what actually happened.

The two case examples illuminate applications that are close to reality of today. The lab-on-a-chip technology is delivering new ideas and methods – some directly applied to separation of molecules. For the large number of applications there, a systematic approach is of benefit.

The other case example, scale-up and adaptation of a downstream operation, may seem as a routine matter. Still, we strongly emphasize that this should be done systematically and not *ad hoc* as is often the case in daily practice.

Perhaps the best advice is to take in all steps in the mechatronic methodology, even a few may lead to faster improvements.

REFERENCES

1. Harrison, R.G., Todd, P., Rudge, S.R., Petrides, D.P. (2003) *Bioseparation Science and Engineering*, Oxford University Press, Oxford.
2. Derelöv, M., Detterfelt, J., Björkman, M., Mandenius, C.F. (2008) Engineering design methodology for biomechatronic products. *Biotechnol. Prog.* 24, 232–244.
3. L., Janson, J. (eds.) (1998) *Protein Purification: Principles, High Resolution Methods and Applications*, 3rd edition (2011) Wiley, New York.
4. Rehm, H.J., Reed, G., Pühler, A., Stadler, P. (1989–1999) *Biotechnology*, 2nd edition, Vols. 1–12, Wiley-VCH Verlag GmbH, Weinheim.
5. *Anonymous Antibody Purification Handbook*, Amersham Pharmacia Biotech, Uppsala, 2000.
6. Sofer, G.K., Nyström, L.E. (1989) *Process Chromatography, A Practical Guide*, Academic Press, London.
7. Peters, M.S., Timmerhaus, K.D., West, R.E. (2003) *Plant Design and Economics for Chemical Engineers*, 5th edition, McGraw-Hill, New York.
8. Atkinson, A.C., Tobias, R.D. (2008) Optimal experimental design in chromatography. *J. Chromatogr. A* 1177, 1–11.
9. Hermanson, G.T., Mallia, A.K., Smith, P.K. (1992) *Immobilized Affinity Ligand Techniques*, Academic Press, San Diego.
10. Roe, S. (2001) *Protein Purification Techniques: A Practical Approach (Practical Approach Series)*, Oxford University Press.
11. Doran, P.M. (1995) *Bioprocess Engineering Principles*, Academic Press, New York.
12. Hage, D.S., Cazes, J. (2006) *Handbook of Affinity Chromatography*, 2nd edition (Chromatographic Science Series), Taylor & Francis.
13. Simpson, R.J. (2004) *Purifying Proteins for Proteomics: A Laboratory Manual*, Cold Spring Harbor Laboratory Press, New York.
14. Robert, K. (1994) *Protein Purification. Principles and Practice Series: Springer Advanced Texts in Chemistry*, 3rd edition, Springer, Berlin.
15. Bailon, P. (2000) *Affinity Chromatography: Methods and Protocols (Methods in Molecular Biology)*, Humana Press.
16. Subramanian, G. (1991) *Preparative and Process-Scale Liquid Chromatography*, Ellis Horwood Ltd, Chichester, UK.
17. Nilsson, M., Harang, V., Bergstrom, M., Ohlson, S., Isaksson, R., Johansson, G. (2004) Determination of protein–ligand affinity constants from direct migration time in capillary electrophoresis. *Electrophoresis* 25, 1829–1836.

18. Schlinge, D., Scherpian, P., Schembecker, G. (2010) Comparison of process concepts for preparative chromatography. *Chem. Eng. Sci.* 65, 5373–5381.
19. Sun, Y., Liu, F.F., Shi, Q.H. (2009) Approaches to high-performance preparative chromatography of proteins. *Adv. Biochem. Eng. Biotechnol.* 113, 217–254.
20. Nagrath, D., Messac, A., Bequette, B.W., Cramer, S.M. (2004) A hybrid model framework for the optimization of preparative chromatographic processes. *Biotechnol. Prog.* 20, 162–178.
21. Levison, P.R. (2003) Large-scale ion-exchange column chromatography of proteins. Comparison of different formats. *J. Chromatogr. B* 790, 17–33.
22. Kaltenbrunner, O., Jungbauer, A., Yamamoto, S. (1997) Prediction of preparative chromatography performance with a very small column. *J. Chromatogr. A* 760, 41–53.
23. Jungbauer, A., Hackl, S., Yamamoto, S. (1994) Calculation of peak profiles in preparative chromatography of biomolecules. *J. Chromatogr. A* 658, 399–406.
24. Jungbauer, A. (1993) Chromatography of biomolecules. *J. Chromatogr.* 639, 3–16.
25. Whitley, R.D., Van Cott, K.E., Wang, N.H.L. (1993) Analysis of nonequilibrium adsorption/desorption kinetics and implications for analytical and preparative chromatography. *Ind. Eng. Chem. Res.* 32, 149–159.
26. Yamamoto, S., Nomura, M., Sano, Y. (1990) Preparative chromatography of proteins: design calculation procedure for gradient and stepwise elution. *J. Chromatogr.* 512, 89–100.
27. Mazsaroff, I., Regnier, F.E. (1986) An economic analysis of performance in preparative chromatography of proteins. *J. Liq. Chromatogr.* 9, 2563–2583.
28. Righezza, M., Chretien, J.R. (1993) Factor analysis of experimental design in chromatography. Part XIV. *Chromatographia* 36, 125–129.
29. Makrodimitris, K., Fernandez, E.J., Woolf, T.B., O'Connell, J.P. (2005) Simulation and experiment of temperature and cosolvent effects in reversed phase chromatography of peptides. *Biotechnol. Prog.* 21, 893–896.

11

Stem Cell Manufacturing

This chapter discusses application of the biomechatronic design methodology for stem cell manufacturing systems. First, the state of the art of stem cell manufacturing is overviewed with emphasis on illuminating the vast complexity of this type of bioprocessing (Section 11.1). The multitude of requirements and needs this advanced technology is accompanied with is discussed and a few examples of target specification tables are shown (Section 11.2). The design methodology is applied with these constraints and various design alternatives are evaluated using the methodology (Section 11.3). Finally, a specific case for human embryonic stem cell (hESC) expansion is shown to exemplify the principles (Section 11.4).

11.1 STATE OF THE ART OF STEM CELL MANUFACTURING

The manufacturing of stem cells is today in its infancy [1,2,3]. It is thus timely to apply established modern methods for design to the systems and processes for manufacturing stem cells and stem cell-derived products and, in particular, to do that with biomechatronic design methodology as a tool [4,5]. This could be a challenging possibility for stem cell R&D companies that intend to

Biomechatronic Design in Biotechnology: A Methodology for Development of Biotechnological Products, First Edition. Carl-Fredrik Mandenius and Mats Björkman.

transform their established cells and cell lines into bioproducts that should be reliably and efficiently manufactured according to current Good Manufacturing Practice (GMP) and in compliance with safety and ethical regulatory directives and guidelines from the European Commission, FDA, EMA, and other regulatory authorities [6–10]. This is especially motivated for clinical-grade products for cell therapy applications where derived organ cell types are to be transplanted into a patient [9,11]. Application of stem cells in the drug discovery and development process must also reach high standards in order to be able to deliver relevant materials in toxicity testing and other preclinical studies [12,13].

The variety of procedural alternatives for propagation, expansion, differentiation, preservation, and characterization of stem cells makes the design task very complex. The need to adapt, refine, and optimize to satisfy the demands of good process economy, high-quality cell products, and regulatory compliance justifies the mobilization of the most powerful design methods.

In the common procedure for producing stem cells materials (e.g., hESC and hESC-derived products), the first step in the preparation is to enzymatically digest the inner cell mass of embryonic blastocytes [14]. The isolated cell mass is spread on a layer of growth-inhibited mouse fibroblast feeder cells in tissue culture dishes. After 1 or 2 weeks, the outgrown cells from the inner cell mass are isolated, by manual dissection or enzymatic treatment, and transferred to new culture dishes for further growth in an undissociated cellular state. The feeder cells provide endogen differentiation inhibitors that hinder the stem cells to differentiate further [2].

The stability of the differentiation state of cell culture is monitored through characterizing specific biomarker molecules, typically cell surface proteins (e.g., SSEA-3/4, Oct3/4, and Nanog), which indicate that the cell mass remains in the undifferentiated state [15]. Additional characterization and control methods could be efficient to use, for example, gene array analyses, flow cytometry, and other methods [2].

If the final product of the process is undifferentiated stem cells, the characterized cells are isolated and further expanded under controlled conditions as seen in Figure 11.1. As much as 100 passages can be attained if the appropriate inhibition factors are added to the process [13].

If the final product is a particular organ cell type, a continuing well-defined differentiation protocol ensues in a stepwise procedure of propagation and isolation of the cells. The added differentiation factors and the culture conditions applied stimulate the cells to develop into progenitor cell states and further to matured cells from which the desired cell type is isolated (see Figure 11.1). This part of the process may take another 2–4 weeks. Very important for a precise control of the development of the cells is to continue characterization of biomarker molecules, now others are characteristic for the

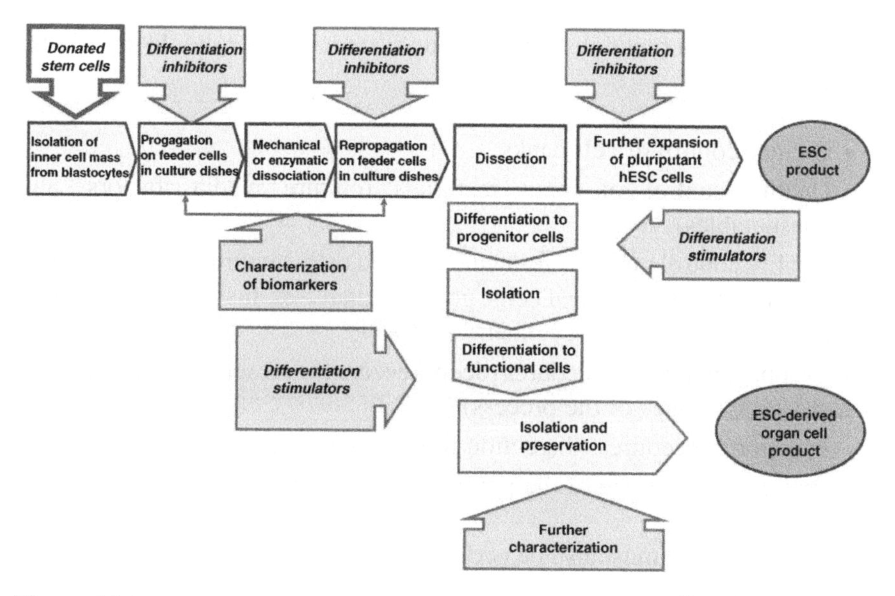

Figure 11.1 *An example of a stem cell production process for hESC and hESC-derived organ cell production. Adapted with permission from Ref [5].*

desired cell type. Other functionality tests can support the biomarker tests to ensure that the right cell type is reached.

During the process, cells go through several isolations by manual microscopic selection and dissection. Between and after the steps, cells are cryopreserved intermittently. The preservation and shipping procedures require careful handling of the cells. Robotic procedures for these steps could be envisaged in future.

The flowchart in Figure 11.1 gives an overview of the steps in derivation of the stem cells and includes in-process analytical characterization.

For full-scale manufacturing, this procedure must be scale up, e.g. with support of bioreactors [16–20]. Now the hESC and hESC-derived products require access to IPR protected cells, perfectly controlled and contamination-free manufacturing facilities, traceability of raw materials, xeno-free media, advanced equipment for handling cell culture, and a variety of analytical instrument and quality control devices. Importantly, very skilled and well-trained staff and operators are needed, or robotics that can take over some of their manual work.

Procedures that can be performed by robotic systems are cell expansion and maintenance of multiple cell lines, various subculturing steps, expanding cell numbers through the seeding of a number of flasks, incubation in different flasks and plate formats, harvesting and plating of cells for assays and screening, cell counting and viability measurement, processing of multilayer flasks, and automated plating [21,22,23].

Other important functions the scaled-up system could or should include are as follows:

- Quality control of cell banks
- Quality control of input materials (culture media, factors, and consumables)
- Quality control methods for manufactured cells (monitoring biomarkers, functional testing of final cells, genomic analyses, histological analyses, microscopy, and flow cytometry methods)
- Cryopreservation procedures (deep freezers for preserving cell materials at various stages of the process)
- Isolation procedures (dissection in microscopy, enzyme treatments)
- Preservation/storage of the product and shipping procedures

The manufacturing must also be performed in clean rooms to reduce the risk for contamination.

All this requires extensive validation and reconsideration of effects caused by the scaled-up manufacturing systems [24].

11.2 NEEDS AND TARGET SPECIFICATIONS FOR SCALED-UP STEM CELL MANUFACTURING

The needs and requirements of a stem cell manufacturing process should in the biomechatronic design methodology we apply be listed and specified with target values.

The List of Target Specifications in Table 11.1 lists important needs as mentioned in the above section. The needs in the table are presented as metrics. A few additional needs with metrics are included in the table, such as use of PAT/QbD [6,25], something that we will elaborate more thoroughly in Chapter 13.

Target values are set, such as the qualitative values (e.g., "Yes/No") or quantitative values (e.g., time and costs for batches and the attained efficiency by automation of the procedures).

The target values given in the table are examples and indicators of what could be typical values in a particular process.

Depending on the type of stem cell manufacturing process, both the final cell product and the starting embryonic stem cell material can vary significantly and must be defined for each process. In Figure 11.2, three alternative process configurations are shown. Consequently, the List of Target Specifications will look different for each of these process configurations.

TABLE 11.1 List of Target Specifications with the Needs and Specifications for Stem Cell Manufacturing Processes

Needs → Metrics	Target Value	Units
Should apply to hESC starting materials	Yes	Yes/no
Materials from different donors shall be used in the same process	Up to 10	No. of donors
The process shall end in mature human organ cell types	Yes	Yes/no
The process shall end in late progenitor cells	No	Yes/no
The process shall be able to start with early progenitor cells	Yes	Yes/no
Each process step shall have critical biomarkers showing differentiation	At least 5	No biomarkers
Rapid sensor methods should be available for biomarkers	90%	% availability
Real-time imaging of cells shall be included as an in-process QC	Yes	Yes/no
Biomarkers (proteins) be analyzed with sensitivity by specific immunosensors	Yes, 5 μg/g cells	Yes/no, detection level
All steps shall be carried out in LAF	Yes	Yes/no
Robots should decrease manual steps	<25%	% manual of op.
Discarded cells low	<15%	% discarded cells
Low price per cell batch		% op. cost of sale
Should apply PAT/QbD	Yes	Yes/no
Should be cGMP adapted	Yes	Yes/no
Total process time moderate	18–25	Days per batch
Process investment returned fast	<6 years	Year of return

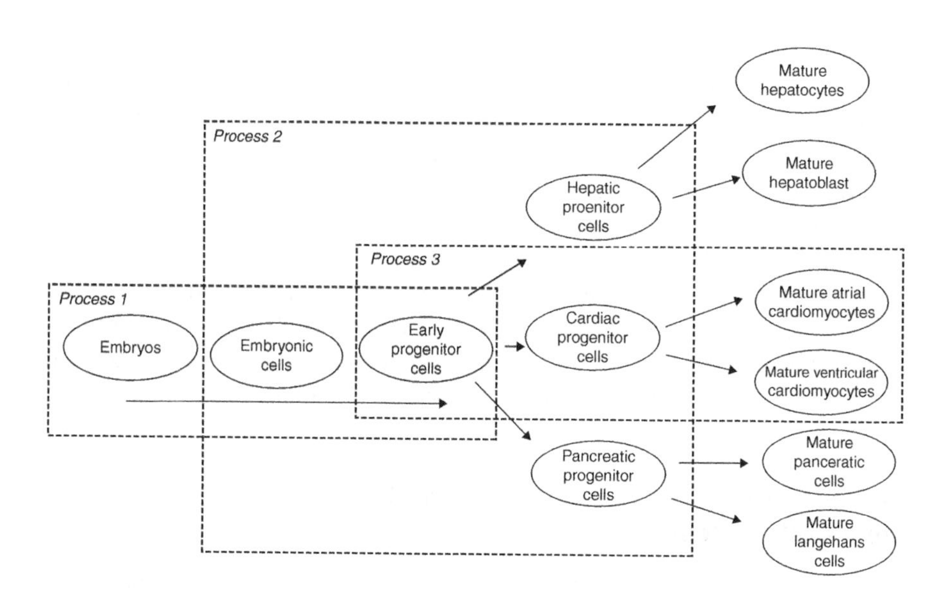

Figure 11.2 *Three stem cell manufacturing process configurations.*

11.3 SETTING UP AN EFFICIENT MANUFACTURING SYSTEM BY USING BIOMECHATRONIC CONCEPTUAL DESIGN

11.3.1 Generating Process Alternatives

The complex technical procedures for stem cell production and the demanding target specifications make it necessary to carefully consider alternatives before going into a primary process flow sheet (i.e., what we in the mechatronic design methodology refer to as the Anatomical Blueprint).

From the general description of stem cell differentiation given above, it follows that there are a number of key issues that must be clarified for being able to generating realistic and useful process alternatives. These issues or elements include the type of target cells that shall be derived from stem cells, if it is an intermediate stem product such as progenitors at different states or a mature organ cell type, the procedures for addition of differentiation factors, what critical biomarkers shall be monitored, which robotic functions should be included, how automatic control functions shall be applied, what degree of containment the manufacturing units require, what other environmental controls are necessary, and which analyzer functions must support the operation of the process. A well-disposed List of Target Specifications should provide most of the conditions and boundaries for these key elements.

On the top of Figure 11.3, there is a Basic Concept Component Chart where these elements are shown and how they at the highest generic level could or ought to be cross-related. The Permutation Chart below has generated three alternatives for configuring the elements: alternatives A, B, and C. In this more detailed chart, the relationships to raw materials (or input operands) and the transformation process (TrP) are configured. The three configurations focus on different alternative control structures. Definitely, many more permutations of the control loops are possible. The shown configurations serve as examples. As could be seen, they also directly relate to the three process system boundaries depicted in Figure 11.2.

It should then be understood that the biomarker function symbol in the permutation chart is not one single biomarker, but preferably the target number given in the specification above (Table 11.1), and the analyzer function should in the same way be understood as all sensors and analyzers that provide the specified information about the cellular state of the transforming cells.

11.3.2 Hubka–Eder Map for a Human Embryonic Stem Cell Process

With these structure alternatives, a Hubka–Eder map is appropriate to establish. As in previous examples, the transformation process is subdivided

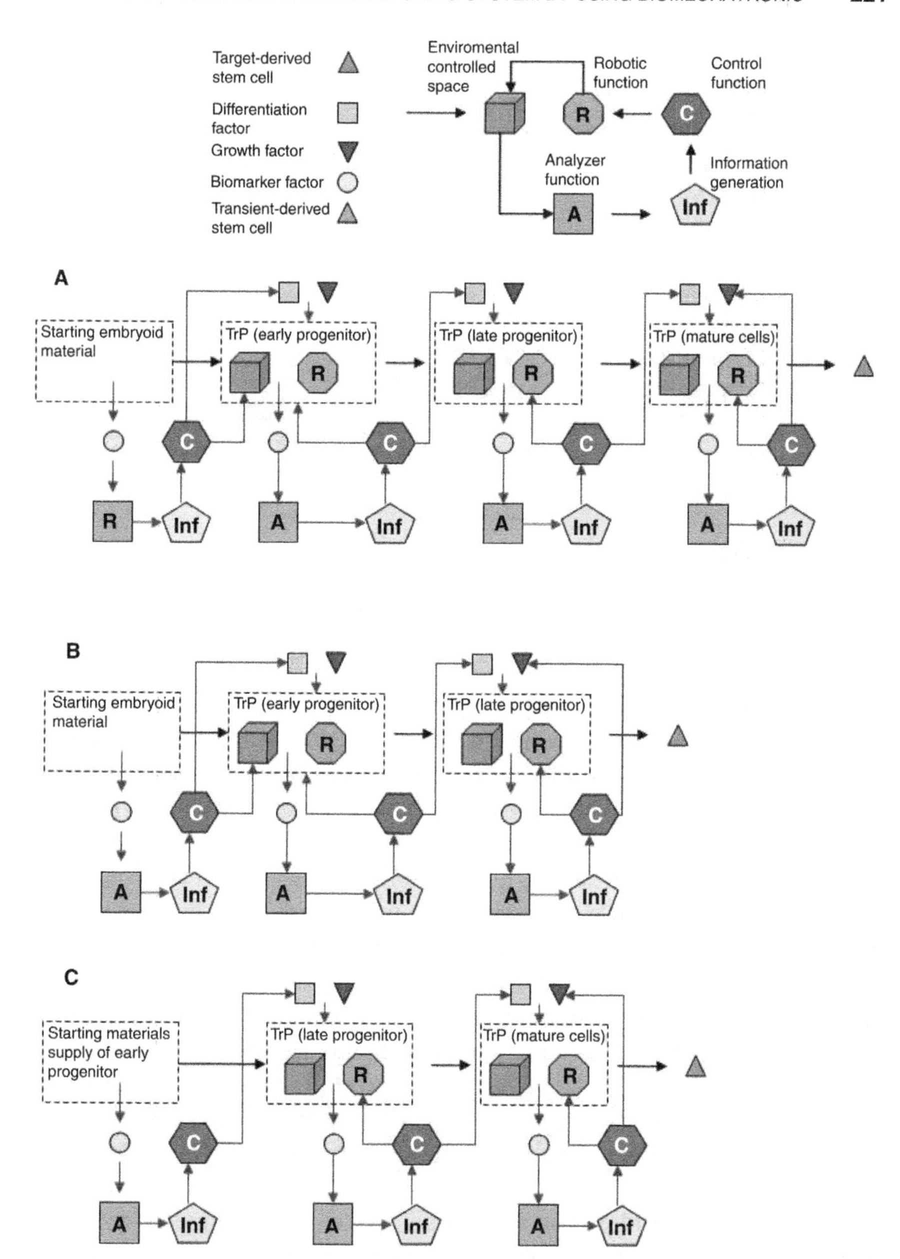

Figure 11.3 *Concept Generation Chart with a Basic Concept Component Chart at the top of the figure and a Permutation Chart with three generated alternatives below.*

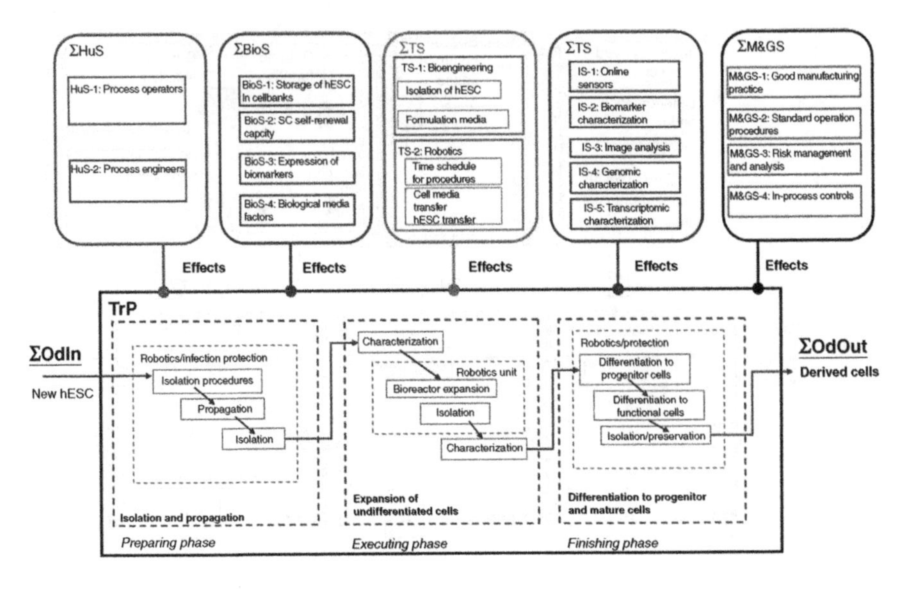

Figure 11.4 *A Hubka–Eder map showing the design of a stem cell manufacturing process. Adapted from Ref. [6].*

into a preparing, executing, and finishing phase (Figure 11.4). The dominating phase is the executing phase, which coincides largely with the flow scheme of Figure 11.1. The preparing phase is the handling of the exclusive starting raw material – the embryonic stem cells – and how they are taken care of in the cell bank. It also includes the pretreatment and preparation of cell media.

The finishing phase structures how to storage the produced target cells and maintaining them until use. It also includes the procedure of shipping the cells to the user (e.g., a clinic or a toxicology test laboratory).

The in- and out-operands are defined as well. The hESC material is the obvious primary input operand together with cell media and growth and differentiation factors. Secondary inputs could be analytical information about cell state, which can adapt the protocol to the TrP or individual customer demands. Out-operands are primarily the derived target cells and characterization data about them. Secondary outputs are residual cells and wasted media.

The ΣBioS is the hESC on a functional level. This includes their functionality in terms of storage ability (BioS-1), their ability of self-renewal (BioS-2), and their expression of characteristic biomarkers (BioS-3). Another important part of the ΣBioS is the functions of the media factors on the differentiation process (BioS-4). Here, we are including a vast number of factors with a variety of biological effects on the stem cells. These functions are, of course, the most essential knowledge embedded in the design of a stem cell process and the result of recent stem cell research and should be detailed to a deeper level in subsystems^n to a degree as relevant for the TrP. This might be a very demanding work and treated here.

Belonging to the ΣBioS is also the basic culture media (BioS-5), normally a combination of biological, organic, and inorganic molecules. Their biological effects on the propagation of the stem cells should be a key function of the ΣBioS entity.

The immunosensors mentioned above need to carry out immunoactive functions for biomarker detection. The system analysis should bring the understanding of the interactions between these process analytical functions and the TrP. Consequently, these functions are also very important to consider as a ΣBioS subsystem.

The ΣTS have many more subsystems than is possible to include in the figure. The complete Hubka–Eder map must however do that.

Important ΣTS subsystems are the functions of the automated handling of liquids by robots (TS-1), the preservation and containment of cells (TS-2), the analytical and imaging devices (TS-2), and the process computer signaling and programs for effectuating the operations (TS-4). All other technical equipment involved in the TrP that have essential functions also become separate technical subsystems. These are the technical subsystems for carrying out functions for centrifugation of cells, incubation, laminar flow containment, microscopic investigation of cells, micromanipulation of stem cells on plates, dispersion, autopipetting, heating, cooling, and so on. These subfunctions are having key effects on the other subsystems and their mutual interactions are very important to capture.

Especially, both the effects that the technical subsystems have on the ΣBioS and the effects caused by ΣBioS on ΣTS should be studied systematically. This includes, of course, both the intended and the unintended effects.

ΣIS include the sensor functions with respect to the information content they create (IS-1), the biomarker information as a fingerprint of the cellular state (IS-2), image analysis data created in, for example, fluorescence microscopes (IS-3), and genomics and transcriptomics data from microarray testing carried out (IS-4).

The ΣIS subsystems are in all cases connected to a technical subsystem that can generate information technically. Other important interactions are with ΣBioS and with databases for data storage.

ΣM&GS include typical process management issues: GMP (M&GS-1), Standard Operation Procedures (M&GS-2), and risk management (M&GS-3), following ICH guidelines. Here, other regulatory practices such as PAT and QbD can also be included as further discussed in Chapter 13.

ΣM&GS have important functions in controlling procedures through automatic controllers, for example, embedded in the robots. The controling is to a large extent a technical subsystem matter, but the choice of set-points or other criteria are a management and goal issues to be handled separate from the technical design of the controllers (M&GS-4).

The ΣM&GS should also include the functions necessary to manage analytical data adequately. This is a typical batch GMP issue that also should be included in the specification and QC protocols to the customers [10].

ΣHuS involved are the skilled operators (HuS-1) and engineers (HuS-2) in the manufacture. Given the high technical level of systems and devices, this becomes a critical function. Good training of personnel requires experienced trainer staff function (HuS-3). The complexity of the process equipment requires a special maintenance and repair function (HuS-4) to be highlighted and analyzed. The support function to provide the customers with adequate information on how to handle the cells is another necessity (HuS-5).

Definitely many unanticipated and uncontrollable influences from an active environment (AEnv) can be envisaged. The biological variation of blood samples in the population is a significant factor to consider. The storage of

Main systems		ΣBioS					ΣTS				ΣIS			ΣM&GS			ΣHuS		
subsystems		Biomarker expression	Storage ability	Diff. factor functions	Basal media functions.	Dedifferentiation	Containment	Robotics	Imaging	Software functions	Sensor data processing	Biomarker data processing	Imaging data processing	GMP compatibility	SOP management	In-process control	Proc. operator capacity	Proc. engineer capacity	Service support
ΣBioS	Biomarker expression		1	1	1	1	1	1	1	1	1	5	3	4	4	4	1	1	1
	Storage ability	1		1	1	1	1	1	1	1	1	5	1	2	2	1	1	1	1
	Diff. factor functions	5	4		1	5	1	1	1	1	1	1	1	2	2	2	1	1	1
	Basal media functions	4	4	3		4	1	1	1	1	3	3	3	2	2	2	1	1	1
	Dedifferentiation	4	4	1	1		1	1	1	1	1	1	1	2	2	2	1	1	1
ΣTS	Containment	4	5	4	4	1		5	3	1	1	1	1	3	2	2	4	5	4
	Robotics	5	5	5	5	5	4		5	4	5	5	5	5	5	2	4	5	5
	Imaging	3	4	4	4	4	5	1		1	3	3	3	3	3	3	4	5	5
	Software functions	1	1	1	1	1	1	5	3		4	4	4	4	2	2	4	5	3
ΣIS	Sensor data processing	1	1	1	1	1	1	1	1	1		1	1	5	5	5	5	5	4
	Biomarker data processing	1	1	1	1	1	1	1	1	4	2		5	5	5	3	4	4	3
	Imaging data processing	1	1	1	1	1	1	1	1	5	5	5		5	3	2	4	4	3
Σ	GMP compatibility	4	4	4	4	4	4	4	4	4	4	4	4		4	4	5	5	4
	SOP management	4	4	4	4	4	4	4	4	4	4	4	4	4		4	5	5	3
	In-process control	1	1	1	1	1	1	1	1	1	1	1	1	5	5		5	5	5
ΣHuS	Process operator capacity	4	4	2	2	5	4	4	1	2	1	5	3	5	3	4		5	3
	Process engineer capacity	1	1	1	1	1	3	3	3	3	1	5	4	5	5	4	5		4
	Service supporting	5	5	5	5	5	2	2	4	2	2	2	2	5	5	5	5	5	

Figure 11.5 *A Functions Interaction Matrix for stem cell manufacturing. The numbers index the strength of the interaction (5 = very strong; 4 = strong; 3 = medium; 2 = weak; 1 = very weak or nonexisting).*

more or less sensitive enzymes and other reagents where climate conditions may affect the shelf life may vary considerably. The climate variation may also influence the performance of the sensor. Sensitivity to humidity and temperature effects on the reaction rates of the analytical conversions need consideration.

The large number of interactions taking place in the complete stem cell manufacturing process motivates the use of a Functions Interaction Matrix.

Figure 11.5 shows an example of the Functions Interaction Matrix for a stem cell process where the main subsystems are included.

As seen, the interactions assessed as very strong and strong are many. Note the unsymmetric character of the interactions. For example, the biological systems have little or no impact on the technical systems whereas the technical systems have in the majority of their interactions significant influence. This could, for example, be that the performance of a robotic subsystem that should handle the cell culture may be decisive for the cells' state after the operation (volume precision and mechanical stress).

11.4 CONCLUSIONS

This chapter has highlighted some of the specific characteristics of hESC manufacture that need extensive attention for a further exploitation of the hESC technology into successful products. Here, a systematic inspired approach can be especially rewarding from both the regulators' perspective and the manufacturers' perspective. To employ the methodology of manufacturing systems design may then be a good way to avoid unnecessary costs and delays in the development.

A manufacturing facility for hESC products should most probably be flexible for production of different final products, for example, different hESC-derived cell types or undifferentiated hESC materials from a particular hESC line. The typical customers will have varying demands depending on the purpose of the produced cell material (for clinical-grade use, for drug testing, etc.). The quantity of the cells required will vary considerably over time. The distribution/shipping conditions will be more or less demanding. A hESC manufacturer may, of course, specialize on certain customers and cell types. However, from a manufacturing engineering perspective, an interesting question is how to make a robust flexible manufacturing system that can cope with the varying demands in order to be as competitive as possible on the market. The demands for repeatability of the production batches imply that robots should be used instead of human operators whenever possible since the variability and risk for mistakes due to monotonous and repetitive work are much smaller for robotic systems than for human operators.

However, many of the inherent criteria of the regulatory demands are manifested in the mechatronic design approach. It is an example of a scientific method and a tool for knowledge management and knowledge analysis, albeit not by using traditional knowledge management tools.

Several features can be envisaged as typical in hESC manufacturing – high demands on robust flexibility, long-time contracts, and very small volumes of bioproducts with an unusually high price. This is in contrast to most other traditional areas where Hubka–Eder mapping is applied.

REFERENCES

1. Sartipy, P., Björquist, P., Strehl, R., Hyllner, J. (2007) The application of human embryonic stem cell technologies to drug discovery. *Drug Discov. Today* 12, 688–699.
2. Ameen, C., Strehl, R., Björquist, P., Lindahl, A., Hyllner, J., Sartipy, P. (2008) Human embryonic stem cells: current technologies and emerging industrial applications. *Crit. Rev. Oncol. Hematol.* 65, 54–80.
3. Adewumi, O., et al. (2007) Characterization of human embryonic stem cell lines by the international stem cell initiative. *Nat. Biotechnol.* 25, 803–816.
4. Derelöv, M., Detterfelt, J., Björkman, M., Mandenius, C.F. (2008) Engineering design methodology for bio-mechatronic products. *Biotechnol. Prog.* 24, 232–244.
5. Mandenius, C.F., Björkman, M. (2010) Mechatronic design principles for biotechnology product development. *Trends Biotechnol.* 28(5), 230–236.
6. Mandenius, C.F., Björkman, M. (2009) Process analytical technology (PAT) and Quality-by-Design (QbD) aspects on stem cell manufacture. *Eur. Pharm. Rev.* 14 (1), 32–37.
7. von Tigerstrom, B.J. (2008) The challenge of regulatory stem cell-based products. *Trends Biotechnol.* 26, 653–658.
8. European Commission (2004) Directive of the European Parliament and of the Council of 31 March 2004 on setting standards of quality and safety for donation, procurement, testing, processing, preservation, storage and distribution of human tissues and cells. E. C. Directive 2004/23.
9. Federal Drug and Food Administration (USA) (2008) Content and Review of Chemistry, Manufacturing, and Control (CMC) Information for Human Somatic Cell Therapy Investigational New Drug Applications (INDs), FDA.
10. Unger, C., Skottman, H., Blomberg, P., Dilber, M.S., Hovatta, O. (2008) Good manufacturing practice and clinical-grade human embryonic stem cell lines. *Hum. Mol. Genet.* 17, R48–R53.
11. Kirouac, D.C., Zandstra, P.W. (2008) The systematic production of cells for cell therapies. *Cell Stem Cell* 3, 369–381.

12. Khetani, S.R., Bhatia, S.N. (2008) Microscale culture of human liver cells for drug development. *Nat. Biotechnol.* 26, 120–127.
13. Jensen, J., Hyllner, J., Björquist, P. (2009) Human embryonic stem cell technology and drug discovery. *J. Cell Physiol.* 219, 513–519.
14. Ström, S., Inzunza, J., Grinnemo, K.H., Holmberg, K., Matalainen, E., Strömberg, A.M., Blennow, E., Hovatta, O. (2007) Mechanical isolation of inner cell mass is effective in derivation of new human embryonic stem cell lines. *Hum. Reprod.* 22, 3051–3058.
15. Chang, K.H., Zandstra, P.W. (2004) Quantitative screening of embryonic stem cell differentiation: endoderm formation as a model. *Biotechnol. Bioeng.* 88, 287–298.
16. Zeilinger, K., et al. (2002) Three-dimensional co-culture of primary human liver cells in bioreactors for in vitro drug studies: effects of the initial cell quality on the long-term maintenance of hepatocyte-specific functions. *Altern. Lab. Anim.* 30, 525–538.
17. Lee, P.J., Hung, P.J., Lee, L.P. (2007) An artificial sinusoid with a microfluidic endothelial-like barrier for primary hepatocyte culture. *Biotechnol. Bioeng.* 97, 1340–1346.
18. Lee, P.J., Hung, P.J., Rao, V.M., Lee, L.P. (2006) Nanoliter scale microbioreactor array for quantitative cell biology. *Biotechnol. Bioeng.* 94, 5–14.
19. Toh, Y.C., Lim, T.C., Tai, D., Xiao, G., van Noort, D., Yu, H. (2009) A microfluidic 3D hepatocyte chip for drug toxicity testing. *Lab Chip* 9, 2026–2035.
20. King, J.A., Miller, W.M. (2007) Bioreactor development for stem cell expansion and controlled differentiation. *Curr. Opin. Chem. Biol.* 11, 394–398.
21. Narkilahti, S., Rajala, K., Pihlajamäki, H., Suuronen, R., Hovatta, O., Skottman, H. (2007) Monitoring and analysis of dynamic growth of human embryonic stem cells, comparison of automated instrumentation and conventional culturing methods. *Biomed. Eng. Online* 6(11), 1–8.
22. The Automation Partnership, Cambridge, UK http://www.automationpartnership.com/
23. Drake, R.A.L., Oakeshott, R.B.S. (2005) Smart cell culture, European Patent Application EP 1598415.
24. Thomas, R.J., Anderson, D., Chandra, A., Smith, N.M., Young, L.E., Williams, D., Denning, C. (2009) Automated, scalable culture of human embryonic stem cells in feeder-free conditions. *Biotechnol. Bioeng.* 102, 1636–1644.
25. Mandenius, C.F., Derelöv, M., Detterfelt, J., Björkman, M. (2007) Process analytical technology and design science. *Eur. Pharm. Rev.* 12, 74–80.

12

Bioartificial Organ-Simulating Devices

This chapter describes the application of the biomechatronic design methodology to bioartificial devices for simulating organs in a body, preferably the human organs. First, the development of these devices during the past 20 years is summarized (Section 12.1). Then, the design methodology is applied to the devices for various applications (Section 12.2). Finally, the device that has attracted most attention, the bioartificial liver, is analyzed in detail (Section 12.3). General conclusions are drawn on how the biomechatronic design methods and tools best can support the design of new organ-simulating devices.

12.1 INTRODUCTION

What is the difference between bioartificial organ devices and the cellular bioreactors described in Chapter 9? The striking difference is, without doubt, that the bioartificial organ devices, often also referred to as bioreactor systems, have quite another purpose of use: instead of producing materials, such as small molecules, proteins, or cells, the artificial organ devices shall replace a living human or animal organ by a mimic that sufficiently recreates the essential functions of the natural organ.

Biomechatronic Design in Biotechnology: A Methodology for Development of Biotechnological Products, First Edition. Carl-Fredrik Mandenius and Mats Björkman.

However, the bioartificial organ devices must not necessarily recreate all of the functions of the natural organ; only those that are needed for the particular applications they are aimed for, for example, replacement therapy, toxicity testing, or extracorporal support.

So far, the majority of the bioartificial organ devices are mimics of the liver and the essential functions of the liver [1–3], while mimics of the kidney and other vital organs have not yet been successfully applied to the same extent [4,5].

One of the most advanced applications, so far, is the use of bioartificial human liver devices in transplantation surgery where it supports the human liver functions for periods of up to 1 day. This may sound as a relatively modest operation time for a bioreactor, but then one must bear in mind that the requirements for its use in clinical practice are significantly more stringent than most other applications [6]. For other bioartificial liver bioreactor applications, for example in toxicity testing, the critical hepatotoxicity pathways are possible to maintain for periods of up to 60 days in scaled down devices [7].

In Figure 12.1, six bioartificial organ-simulating devices for liver or kidney cells are shown that illuminate the variation of design solutions that have unfolded during the past years.

One of these systems, a bioartificial liver support device developed at Charité Universitätsmedizin Berlin, has successfully been applied in transplantation surgery [6]. The system is commercially available since 2000 (Stem Cell Systems GmbH, Berlin, Germany).

Most of the bioartificial liver systems can be categorized as small-scale liver prototypes with slightly varying designs based on multicompartment concepts. Several of these have been developed by Bader and coworkers at the University of Leipzig [8,9]. These multicompartment bioreactors have also been demonstrated with other organ tissues [9].

Microscale mimics of liver [10,11] and kidney [12,13] have, so far, only been realized in proof-of-concept research. Especially, the integrated "lab-on-a-chip" configurations succeed in recreating many of the intricate microstructures of the human liver due to the ability of the chips to miniaturize channels to dimensions on the scale of capillaries [10].

The recreation of the 3D architecture of the organ tissue with various scaffolds has become a key element of the design, which has been elaborated by many research groups [14]. Different biomaterials have been investigated such as alginate hydrogels, biocompatible organic polymers, and various natural materials. The driving design principle has been to access a porous microstructure that imitates typical dimensions and distances in the tissue; between different cell types, between ducts and capillaries, or other cellular structures that form the tissue. Recently, nanobiotechnology and microlithographic techniques have been exploited (cf. Figure 12.4).

Bioartificial organ device	Principle	Reference
Radial flow bioreactor 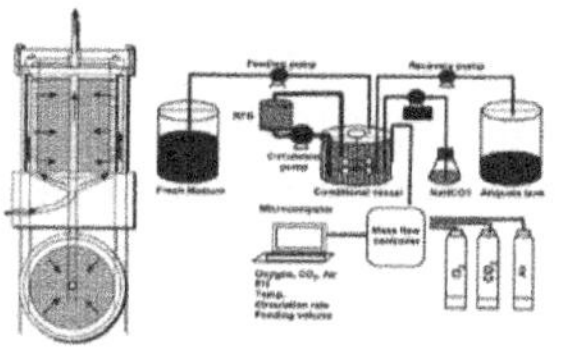	A three-dimensional perfusion culture of hepatocellular renal cells, to create a bioartificial liver and kidney. The cylindrical reactor is filled with porous cellulose microcarrier. The medium flows from the periphery toward the center, delivering supply of oxygen and nutrients to cells at the center as well as at the periphery.	[15]
Oxygen-permeable membrane bioreactor 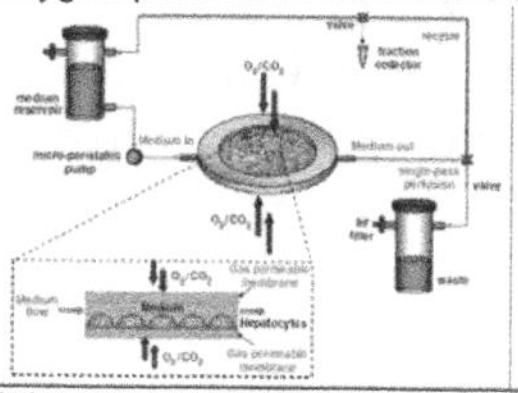	Cells cultured between flat-sheet gas-permeable polymeric membranes, ensuring the diffusion of O_2 and CO_2 and providing a support for cell anchorage and growth. Permit online observation of the cells with inverse microscope.	[8]
Silicon-based device 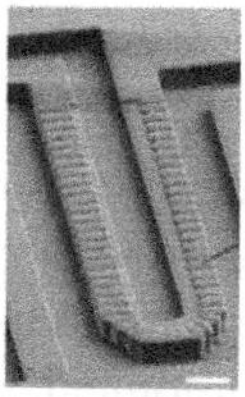	A silicon-based microfabricated device that mimics the capillary system of the liver tissue. The physiological milieu of a liver tissue is recreated within the silicon microstructure.	[10]
Four-compartment artificial liver 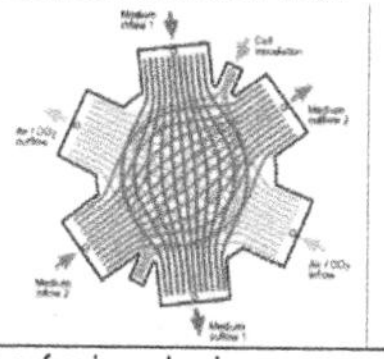	Four-compartment liver tissue bioreactor: Capillaries are interwoven, forming a tight tissues-like network with three compartments. Liver cells in a 3-D microstructure are filling up the space in between (forth compartment).	[16]
Analytical perfusion device 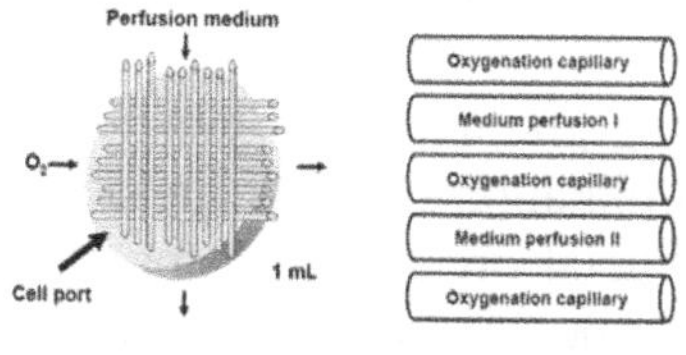 Perfusion medium O_2 Cell port 1 mL Oxygenation capillary Medium perfusion I Oxygenation capillary Medium perfusion II Oxygenation capillary	Analytical scale 3-D perfusion device with four separate chambers connected to same stream. Counterdirectional flow via two independent capillary compartments over hydrophilic membrane bundles, mixed gas for oxygenation and CO_2 perfusion by hydrophobic capillary membranes	[7]

Figure 12.1 *Examples of design solution for bioartificial organ devices. Reproduced with permission from [8], [10], [15], and [16].*

Bioartificial kidney devices could serve as replacements for hemodialysis with continuous hemofiltration for providing fully controlled metabolic functions of the renal tubules [4]. These devices simulate the renal tubular epithelial cells. Active transport is created on membrane structures in confluent monolayers. Clinical targets for use are renal diseases. Problems encountered for good functional properties of the devices are antithrombogenic properties, and transport capacities of the tubular system. Connections of hemofilter devices with different membranes have been tested in various setups with a varying degree of success [17–19]. Attempts to realize devices with renal stem cells/progenitor cells with human origin and with higher relevance are new avenues under investigation [13].

In this chapter, we show how the biomechatronic design methodology can be applied to human organ-simulating devices, with preference for the human liver. Two examples of existing designs are analyzed in detail: a four-compartment bioartificial liver bioreactor (from the Charité Universitätsmedizin Berlin, Germany) [1,3,6] and a perfusion bioreactor for primary liver cell (from the University of Leipzig, Germany) [20,21].

12.2 DESIGN OF BIOARTIFICIAL ORGAN-SIMULATION DEVICES

The following section describes the basic requirements for the organ-simulating devices using the biomechatronic design methodology and design tools as outlined in Chapter 4. The main purpose here is to set the constraints for the design issues and relate these to the biological and medical expectations.

12.2.1 Needs and Specifications

As described in the Section 12.1, bioartificial organ-simulating devices are a heterogeneous group of constructions with a wide variation in requirements. Especially in this case, it seems motivated to detail the needs and target specifications of the design.

The purpose of the design and the ensuing product development for an artificial organ device must begin with the formulation of a clear mission. For example, this could be to develop a bioartificial organ device for clinical applications or to develop a bioartificial liver device for testing pharmacological effects of a new drug *in vitro*.

The needs of the users may then include specific needs related to critical conditions for surgery, or specific critical needs for pharmacological assessments of the drug, such as safety and efficacy aspects, *in vitro* toxicity testing such as acute and repeated toxicity, or organ-specific toxicity.

For liver devices, several needs are common for all users while some are specifically related to the application [22–27]. Drug testing, for example, requires multiple tests and high reproducibility and repeated toxicity testing requires homeostasis of cells. Maintaining functional hepatic cells with the desired phenotype is a particularly important feature for drug evaluation and a necessity for meeting regulatory requirements.

The varying needs may generate the question if a bioartificial organ platform product should be designed that could meet all three main applications, or, if instead three separate stand-alone devices should be designed.

The different user needs are provided with target specifications. In the List of Target Specifications in Table 12.1, the target specifications are either qualitative metrics with a discrete value or quantitative metrics with a discrete value, an upper or a lower limit, or a defined range [28].

Identification of appropriate metrics depends on application and organ and different potential users have at this stage of development diverse opinions about what are the necessary levels.

Sometimes, the target values are optimistic wishes rather than realistic goals. This is especially true for some of the hepatocyte properties, such as the longevity of the liver cells' homeostasis.

The List of Target Specifications given in Table 12.1 sets the values that we will use in the subsequent conceptual analysis. It is based on experience from our own data and on literature studies. However, every new design task requires a new set of values.

From the target specifications, several conceptual design solutions can more or less directly be generated. Several of these have already been realized previously. The Concept Generation Chart of Figure 12.2 depicts six such concepts. These are chosen to illuminate characteristic design concepts that exhibit diverse technical solutions that are stringently different in design in order to show the methodology (Figure 12.2). Thus, it should be realized that several other conceptual designs also could be generated. It should also be noted that the concept generation should not bring ready design architectures showing the physical components of the concept. Instead, the functions should be exposed allowing further elaboration and combination.

The upper part of Figure 12.2 shows a Basic Concept Component Chart where the functions of the organ bioreactor are described from its most generic components. These shall be possible to permute or recombine in a number of inventive configurations. The lower part of the figure, the Permutation Chart, shows the six alternatives that were generated.

Alternative A is a concept based on more or less a conventional lab-scale bioreactor with the hepatocytes grown in particulate 3D scaffold structures. The media (nutrients, gases) are perfused through the reactor countercurrent and the environment in the reactor is controlled as in a lab-scale reactor with sensors for pH, dissolved oxygen, and temperature. Previous studies have demonstrated similar devices based on this concept using, for example, porous microcarriers [29] or cage-like containers [15] used as support for hepatocyte cultures.

TABLE 12.1 User Needs and Target Specifications for a Bioartificial Organ Device

Needs → Metrics	Target Value	Units
Performance needs		
High stability	Should keep stable operation 1 week	Week
High measurement sensitivity	Should respond to toxicity effects of less than 1 μg	μg/100 hepatocytes
Short analysis response time	Time between exposure and detectable response	Minutes
Low detection limit	Viability threshold	
High precision for analysis of crude biological samples		
High throughput of samples	Should manage 10 samples per hour	Number of samples
3D architecture duration	10 days	Days
Endothelial barrier mimic	Within 90%	Mimic degree
High accuracy for analysis of crude biological samples	Blood sample	Sample character
Convenience needs		
Convenient operation temperature	15–30°C	°C
One-person operation	Analysis should not require more personnel force	Y/N
Should be QC/QA adapted	Should meet FDA requirements	Y/N
Need for lab	GLP lab standard	
Size of instrument	Bench scale system	
User-friendly software	1-day training needs	Training time in days
Technical–operational needs		
Biological operation pH	Device pH range between 6 and 8	pH units
Temperature of operation	15–30°C	°C
Small sample volume	Lowest substance amount detected	μg/100 hepatocytes
Multiplex sample exposure	10–20 parallel systems	Parallel sample no.
Perfusion flow	100 pL/(s unit)	Volume/s and unit
Reynolds number in tissue unit	<0.01	Dimensionless
Peclet number in tissue unit	Between 0.8 and 56	Dimensionless
Capacity needs		
High throughput of samples	Should manage 10 samples per hour	Number of samples
Other criteria		
Low price per sample	50 EUR/sample	EUR/sample
Hardware support provided in short time	3 days	Days
Consumables delivered within short time	2 days	Days

In alternative B, a hollow-fiber cartridge-type column is used for separating the cell culture from the component and perfused blood and/or media components in order to protect the cells, providing a surface for cells to exert their function.

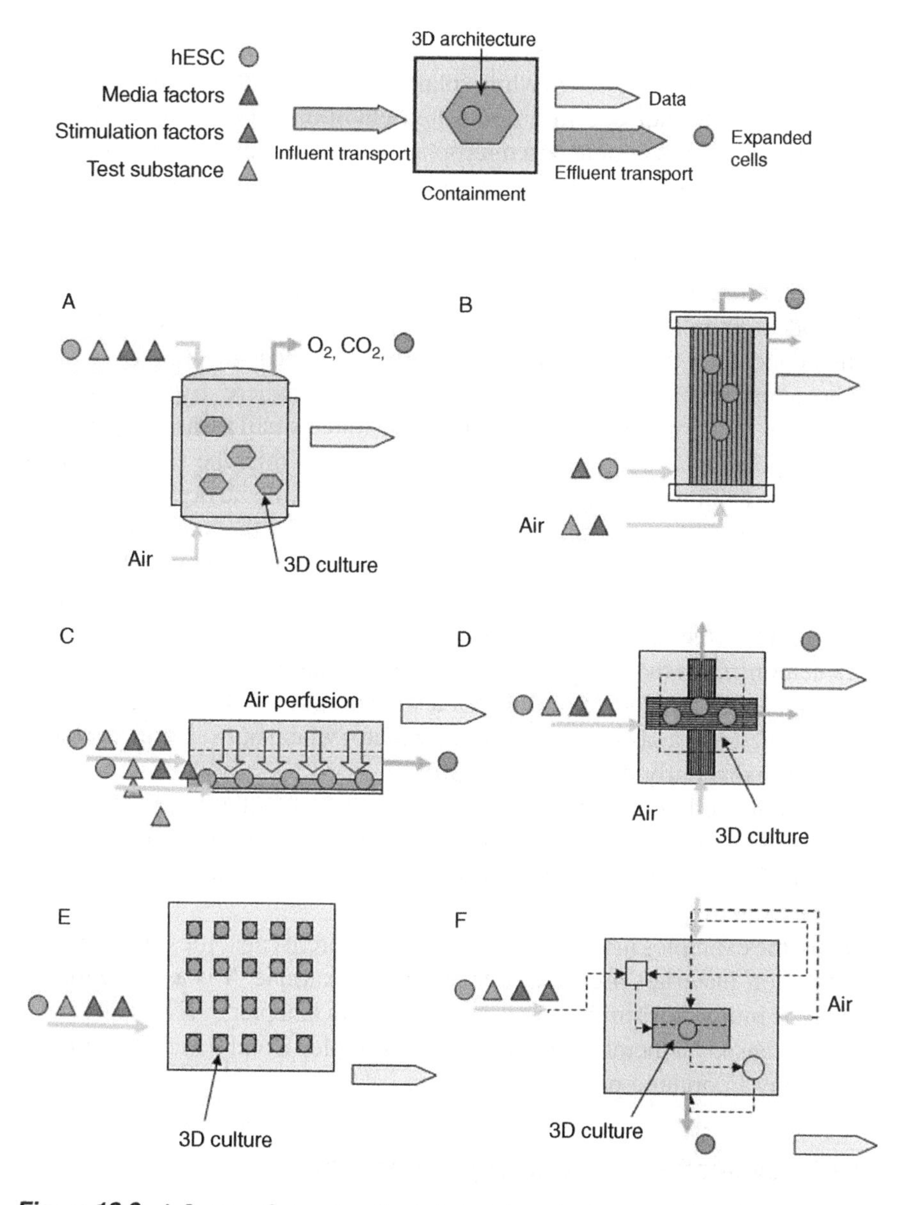

Figure 12.2 *A Concept Generation Chart derived from the target specification: (A) a stirred bioreactor with encapsulated or aggregated 3D cell structures, (B) a hollow-fiber bioreactor where cells are compartmentalized inside the fibers, (C) a trans-well device with oxygenation and nutrient supply through perfusion and micropermeable membrane, (D) an interwoven hollow-fiber perfusion device where tissue cells are on the extracapillary side of the fibers, (E) a microarray-like chip with 3D immobilized cell tissues on the spots of the array, and (F) a lab-on-a-chip device with microfluidic circuits engraved on silicon wafer with a spin coat of silicone/acrylic layers.*

In alternative C, a much simpler device is generated by letting cells grow on a membrane and be oxygenated while small open vessels contain the liver cells and nutrients, all enclosed in a controlled environment (a trans-well). For example, the container could be a microplate with a gas-permeable membrane at the bottom of each well and placed an incubator as demonstrated by Schmitmeier et al. [30].

The hollow-fiber concept is furthered into the alternative D where a multicompartment reactor device with the gas and liquid phase are separated independent of cell hepatocyte culture. A well-known device designed based on this concept is the Charité four-compartment bioreactor [1,3].

The alternative E concept is an array device where liver cells with appropriate functionality are dispersed on separated small areas and perfused with components as addressed by microfluidics. By this, the required functions for the testing purpose are achieved. In this system, the organ cells could be engineered to express functional genes relevant for the application.

If all functional components can be housed in the same block of suitable materials, the separate functional parts above may be realized in a microfluidic plastic unit that then can be significantly reduced in size (alternative F). Recent work demonstrates that this is possible and allows quite small dimensions and ability to handle hepatocyte cultures for testing drugs [11]. The design concept can be refined into a sandwich-membrane reactor where nutrients are perfused by pumps as recently demonstrated [30].

A separate cell-containing unit with the purpose of supplying a user with ready-to-use cells in a single-use unit is shown in several of the concepts in Figure 12.2. The single-use unit is docked with a permanent unit with all necessary control functions.

Thus, the examples in the figure have quite different capacities and different construction materials are used. For example, Example A is a conventional glass-type bioreactor similar to those described in Chapter 9. Examples B, C, and D are based on membrane operations. Examples E and F make use of soft lithography in organic polymer materials.

12.2.2 Evaluation of the Design Concepts

Based on the target specifications, the six concepts are screened (Table 12.2) and scored (Table 12.3). A selection of the target specifications from Table 12.1 are accounted for in Table 12.2 where the six concepts are compared at three levels (− ; 0; +). The matrix screening scores of the concepts single out the *Perfusion* and *Lab-chip* concepts. The scoring also indicates that the microarray concept is relatively acceptable. These three concepts are therefore chosen for further testing. It should be stressed that at this stage the scoring is mostly based on literature data and few experimental

TABLE 12.2 Concept Screening Matrix for Selection Criteria for Bioartificial Devices

Selection Criteria	Concepts					
	Concept Scaffold	Concept Hollow-Fiber	Concept Trans-Well	Concept Perfusion	Concept Microarray	Concept Lab-Chip
Functionality						
Lightweight	−	−	0	0	+	+
In situ monitoring	+	+	−	+	−	+
Online control	+	+	−	+	−	0
3D tissue microstructure mimic	+	+	0	+	−	+
Oxygenation	+	+	+	+	−	+
CO_2 removal	+	+	+	+	+	+
Nutrient supply	+	+	−	+	−	0
Controlled perfusion	−	+	−	+	−	+
Convenience						
Number of units in system	−	−	0	−	+	+
Time for testing	0	0	+	0	+	+
Preparation time	−	−	0	−	+	+
Spare part availability	−	−	0	0	0	0
Ergonomics						
Detector unit	0	0	0	0	0	0
Priming culture infusion	−	−	−	−	−	−
Disposal unit	−	−	+	+	+	+
Durability						
Storage of single-use unit	0	0	+	+	+	+
Biocompatible materials	0	0	0	+	−	−
Shortest runtime of hardware	0	0	0	0	0	0
Total runtime	0	0	0	0	0	0
Performance						
Available runtime per test	−	−	0	+	+	+
Sensitivity of online sensors	+	+	−	+	−	0
Cell volume capacity	+	+	0	+	−	−
Repeatability	+	+	−	+	−	+
Start-up time	−	−	−	−	+	+
Cost						
Total system cost	−	−	+	+	+	0
Single-use unit cost	−	−	+	+	+	+
Operator cost	−	−	−	+	+	+
Spare parts	−	−	+	+	+	+
Sum +'s	9	10	8	18	13	17
Sum 0's	6	6	11	6	4	8
Sum −'s	12	12	9	4	10	4
Net score	−3	−2	−1	14	3	13
Rank	6	5	4	1	3	2

TABLE 12.3 Concept Scoring Matrix for Selection Criteria

		Concept Perfusion		Concept Microarray		Concept Lab-Chip	
Selection Criteria	Weight	Rating	Weighted Score	Rating	Weighted Score	Rating	Weighted Score
Flexible use	10						
Toxicity testing lab	4	4	16	8	32	8	32
Research lab	4	7	28	5	20	5	20
Clinic	2	7	14	2	4	2	4
Mimicking	20						
Tissue scaling factor	10	7	70	1	10	6	60
Oxygen factor	10	7	70	1	10	6	60
Performance	25						
Test time	4	2	8	6	24	6	24
Sensitivity of sensors	4	7	28	1	4	7	28
Repeatability	4	7	28	3	12	6	24
Cells per unit	4	8	32	2	8	2	8
Run time limit	4	8	32	1	4	7	28
Contamination limit	5	7	35	1	5	8	40
User convenience	15						
Easy training	5	3	15	6	30	3	15
Support	5	5	25	6	30	5	25
Test preparation	5	4	20	4	20	4	20
Cost	10						
Investment	3	7	21	2	6	6	18
Consumable cost/test	4	3	12	6	24	4	16
Operation cost/test	3	3	9	4	12	6	18
Manufacturability	20	4	80	7	140	7	140
Total score			543		395		580
Rank			2		3		1

tests since device prototyping is costly and time consuming. In the next round of selection, a much more thorough evaluation is done. Preferably, the three selected concepts are prototyped in smaller lab versions and run with comparable experimental plans where metrics can be generated and reproduced. In Table 12.3, this is exemplified for a slightly modified list of selection criteria. The attributes and targeting have here been refined in order to fine-tune the evaluation and adapt it to the conditions of the experimental plan.

In the Concept Scoring Matrix, eight levels are assessed for the concept and weights are introduced for the importance of the needs. Ratings and weighted scores are balanced based on designer decisions. No doubt this introduces a bias in the evaluation, which can motivate a reiteration later in the development work. It should also contribute to draw attention to redesign possibilities and should facilitate to identify weaknesses in the concepts.

The selection criteria are slightly modified. The total scoring of the three concepts gives again clear preference to perfusion and lab-chip concepts versus the microarray concept.

Thus, these two concepts are further developed, prototyped, and analyzed with Hubka–Eder mapping.

12.3 ANALYSIS OF BIOARTIFICIAL LIVER SYSTEMS

Two of the highlighted design solutions above are here further analyzed by applying a Hubka–Eder mapping [31,32].

The transformation process with in- and out-transport is consequently restricted to the influent blood/nutrients/test compounds, gasses the conversions carried out by the hepatocytes/coculture and the contacting transfer of these parts.

Also included in the TrP process is the generation of signals, physical (e.g., heat, vision), chemical (e.g., metabolites), or biological (e.g., protein release).

The purpose of the model, to analyze the interaction between the systems and the parts thereof, follows.

12.3.1 Biological Systems

The functionalities of the biological systems (ΣBioS) of the device vary considerably depending on application. If toxicity testing is the purpose, the desired functionality of the cells is to carry out by the metabolic conversions that the cell type is specialized in. However, for the two conceptual designs discussed here, the acting ΣBioS have the same functional properties. Thus, the same description of the ΣBioS becomes necessary [33].

Especially for the liver, this is thoroughly described in the toxicology science. The phase I and II enzymes for conversion of xenobiotic compounds are typically composed of three to eight enzymatic steps. Transferases conjunct the xenobiotics to adducts and transporter proteins excrete these from the cell. This chain of enzymatic steps exerts the functionality. The expression mechanisms for creating the functionality are both partly known and partly unknown. Control of the liver functionality is possible through epigenetic factors and other differentiation factors.

Cell–cell communication is believed to have a key role in the development of the functional state of the liver cell. Cocultures of liver cells that balance phenotypic cells favor this development. These cocultures require a scaffold that can recreate the cellular tissue structure.

Thus, the ΣBioS of a bioartificial organ-simulating device of a human liver should have these functional capacities: the ΣBioS should provide phenotypic hepatocytes (BioS-1), allow propagation on cocultured feeder cells (BioS-7), exhibit a 3D architecture realized by a biocompatible microstructure (BioS-7), and should provide a biochemical microenvironment that furnishes the

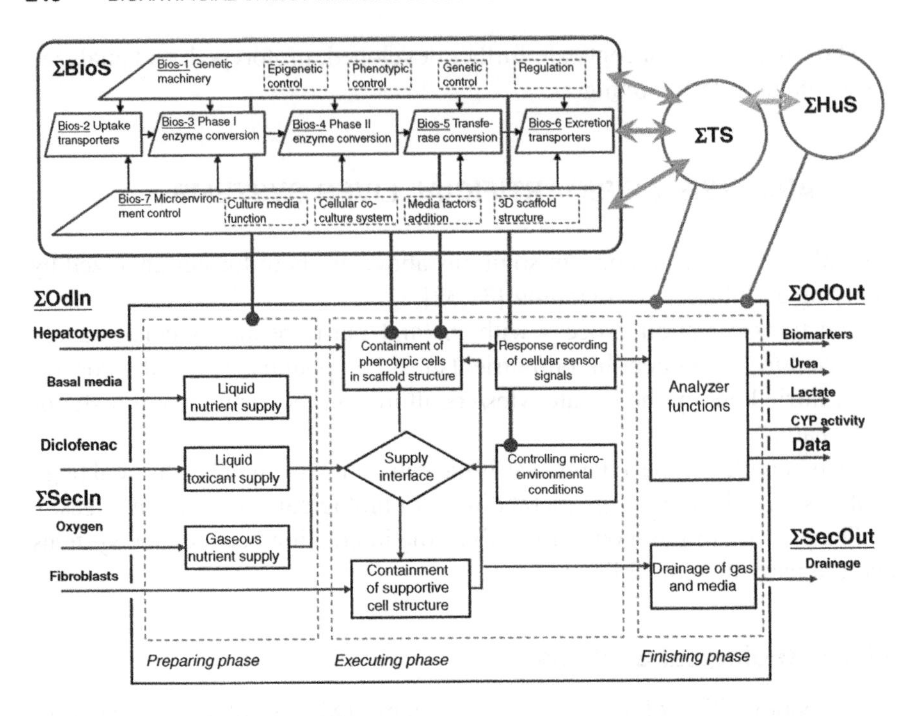

Figure 12.3 *A Hubka–Eder map for the biological systems and transformations of the bioartificial liver device applicable for the lab-on-a-chip and the perfusion concepts. The application is toxicity testing (diclofenac is chosen as example).*

stimulating factors that mature the cells to exhibit the xenobiotic degrading capacity (BioS-7).

Figure 12.3 displays the most important interactions of the ΣBioS with the transformation process of a liver device using a Hubka–Eder map.

Similar Hubka–Eder maps can be set up for other organ cells.

A consequence of the mapping methodology is that the *functions* of the metabolic and physiological machinery of the liver should be detailed in the Hubka–Eder map. Here, it is tempting to just redraw typical charts from biochemical and anatomical text books. That is not the way to do it! Instead the key anatomical and physiological functions of the liver should be carefully represented in the following aspects: the liver's function for adsorbing xenobiotics (BioS-2), for converting the xenobiotics into charged metabolic products in steps (i.e., phases I, II, and III of the liver metabolism) (BioS-3, BioS-4, and BioS-5), and for the excretion of these charged metabolites from the cell (by transporter proteins) (BioS-6). Eventually, the genetic machinery of the liver to express the CYP enzymes could be included (BioS-1) as well as the duct and channel system of the tissue.

The idea of this procedure is mainly only one: to make it possible to test how the functions of other systems may interact and impose effects on the biological systems of the liver and vice versa. This should lead to a systematic investigation of the effect of especially the ΣTS effects.

12.3.2 Technical Systems

Perfusion of media components and test substances through the tissue structure are the key transformation processes of the 3D design of the technical systems (ΣTS) for the perfusion and lab-on-a-chip concepts. The technical solutions for realizing this are constructions with membranes that separate the blood and bile sides from the tissue cells. Thus, the membranes constitute the technical component responsible for infusing the nutrient solution or gaseous constituents from the perfusate flow. The membrane components should then provide a hydrophilic or hydrophobic interface toward the transfusants and a diffusion resistance or permeability that mimic the natural organ counterpart. These barriers provide a 3D architecture that compartmentalizes the design of the device.

The technical systems of the design solution should primarily realize the 3D architecture of the liver tissue. This encompasses creating and mimicking the endothelial barrier on the liver sinusoid. Key parameters are geometrical dimensions and flow properties. These can be expressed from flow rate metrics, cell density, packing degree, dimensionless numbers for the flow, and convection/diffusion ratio (i.e., Peclet number), the laminar flow in the microcirculation units, or other parameters [10].

The devices should most probably also be multiplexed to permit parallel analysis.

For realizing the ΣTS there are many alternatives. Attractive alternatives are the flat membrane systems, hollow-fiber systems, application of encapsulation technology, and the use of aggregated tissue cells or spheroids [2]. The alternatives outlined in Figure 12.2 and in the six generated concepts cover these design solutions.

Figure 12.4 exemplifies another design possibility, to mimic the structure in microfluidic devices, based on lab-on-a-chip fabricated structures. By creating microconduits in silicon/silicon dioxide using the photolithographic methods, structures and dimensions closer to the biological tissue can be mended. The fabrication of integrated circuits also favors the mass production process. Although the channel system is relatively easy to recreate, it is a bigger challenge to succeed with actuator units in the same scale.

Consequently, the lab-on-a-chip concept alternative will require a significantly different set of functions due to the smaller scale. The perfusion concept

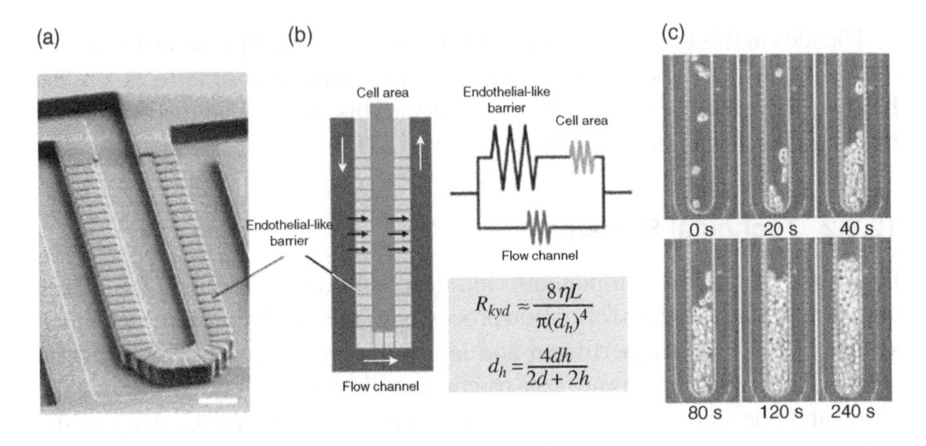

Figure 12.4 *Microfluidic endothelial-like barriers created in silicon structures that mimic the liver capillary flow systems. Dimensions come close to the human tissue liver with channels of 30 μm width and 2 μm height. Adapted with permission from Ref. [10].*

relies on the more established membrane technology with functions better characterized.

As shown for other examples in this book, the ΣTS of the bioartificial liver devices must also include functions effectuated by pumps, valves, containers, or other hardware components.

Supporting mathematical engineering models could be incorporated into descriptions of the ΣTS, for example, by relating to the target specification parameters. This could help to more directly establish state variables of performance of biological conversions and transport processes.

12.3.3 Information Systems

Information systems (ΣIS) required in the devices concern

- monitoring of basic operational parameters (temperature, pressure, flow rates, pumps, and valves);
- monitoring of operational parameters for controlling the state of the cell culture (oxygenation level, nutrient supply, perfusion rate, albumin excretion, LDH release, urea and lactate release, biomarker expression, morphology);
- measurement of toxicological response (partly the same as above but with other criteria of detection concerning sensitivity, response, etc.)

Some of the information is required in real time to be able to induce actuator actions. Others can cope with delayed responses. In principle, the toxicological response allows computing with models and data processing for statistical analysis comparison with databases and other modeling efforts for evaluation.

Analytical systems for monitoring and control are also to be integrated to enable evaluation of cellular responses caused by the exposure of toxicants and to control the unit during operation. Optical methods, possible to implement, such as *in situ* microscopy could be used (e.g., phase contrast and fluorescence microscopes, probes for immunoanalysis, and fixative agents).

ΣIS include functions for sensor signal processing and imaging systems for observing the state of the cells and offline analytics.

As a consequence, data fusion can provide further evidence of the toxicological response and be an integral part of the ΣIS.

In a deeper analysis, the Hubka–Eder map could here also amalgamate with traditional mathematical bioengineering models at various levels, in the calculation in the ΣIS.

12.3.4 Management and Goals Systems

Management and goal systems (ΣM&GS) include the action and decisions taken by the operators and the automatic control actions carried out by the ΣTS actuators. Rules and guidelines for Good Laboratory Practice and Good Clinical Practice can be applied when appropriate. Depending on the purpose of data analysis, the level of data management and adherence to guidelines are set. Organizations such as European Centre for the Validation of Alternative Methods (ECVAM) and equivalent devise guidelines for testing and validations that contribute to ascertain quality of use of the devices. These requirements should be considered when designing the systems.

12.3.5 Human Systems

Human systems (ΣHuS) are involved in several respects. In case the device is used in the clinic, the ΣHuS include the patient and the hepatocyte.

Patients put under liver support systems are key interactive partners who carry demands on the devices.

Surgeons and nurses using the devices during transplantation surgery should operate and interact with these in a safe and reliable manner.

Drug developers and toxicologist to be using the data generated and to set the criteria for the operation and experimental planning must be involved in the design of the systems to accomplish a user-adapted device.

Also, the AEn represents the unpredicted behavior of the patient, variation in the hepatocyte, and unanticipated disturbances in the ambience of the surgery.

Additional subdiagrams detailing ΣHuS as for ΣBioS as in Figure 10.3 would further improve and expand the analysis with additional refinements and testing of user preferences. With this description of the system functions, an anatomical depiction could be done.

12.4 CONCLUSIONS

In this chapter, we have applied the biomechatronic design methodology to a more complicated biological system than in the previous chapters—to a whole human organ, the liver. Still, the design principles are applied in the same way. It shows the strength and flexibility of the mechatronic design tools; they can be applied to any engineering system with the guiding principle of describing the functions and their interactions with the transformations. It is applicable to the bioartificial liver bioreactor for clinical or toxicology use, or any other organ. For example, the working scheme can be applied similarly to human kidney [30], cardiac cells [34], and pancreas as well as neural cell systems of the brain and can be modeled according to the same principles.

Despite the complexity of a human organ, it is relatively simple to rearrange the concepts into the modified established mechatronic models also for a biomedical application. As long as the biological knowledge exists and is accessible, it is also possible to represent it in the mechatronic models.

Two alternative product uses have been mentioned. The analysis has shown that the design architectures had in common many of the user-needs either for clinic or *in vitro* toxicity testing.

A crucial discriminator for the selection of the concept is if the product should be delivered to the customer ready for use with cells and media or if the device should be a tool with which the user itself shall provide hepatocytes or cocultures, differentiation media, and so on, or combinations thereof. This definitely puts different demands on the product design and its manufacture.

In the design solution, it is of value to also consider manufacturing benefits/costs in the concept screening phase. The benefit of manufacturing in lesser number of subunits should be evaluated and, if convenient, an architecture with fewer steps be used.

The mechatronic methodology intends not only speed up the development process but also increase the quality of design solution at early stage and avoid flaws.

REFERENCES

1. Gerlach, J.C. (1996) Development of a hybrid liver support system. *Int. J. Artif. Organs* 19, 645–654.

2. Diekmann, S., Bader, A., Schmitmeier, S. (2006) Present and future developments in hepatic tissue engineering for liver support systems: state of the art and future developments of hepatic cell culture techniques for the use in liver support systems, *Cytotechnology* 50, 163–179.
3. Zeilinger, K., Holland, G., Sauer, I.M., Efimova, E., Kardassis, D., Obermayer, N., Liu, M., Neuhaus, P., Gerlach, J.C. (2004) Time course of primary liver cell reorganization in three-dimensional high-density bioreactors for extracorporeal liver support: an immunohistochemical and ultrastructural study. *Tissue Eng.* 10, 1113–1124.
4. Saito, A., Aung, T., Sekiguchi, K., Sato, Y., Vu, D.M., Inagaki, M., Kanai, G., Tanaka, R., Suzuki, H., Kakuta, T. (2006) Status and perspective of bioartificial kidneys. *J. Artif. Organs* 9, 130–135.
5. Minuth, W.W., Strehl, R., Schumacher, K. (2004) Tissue factory, conceptual design of a modular system for the *in vitro* generation of functional tissues. *Tissue Eng.* 10, 285–294.
6. Gerlach, J.C., Zeilinger, K., Sauer, I.M., Mieder, T., Naumann, G., Grnwald, A., Pless, G., Holland, G., Mas, A., Vienken, J., Neuhaus, P (2002) Extracorporeal liver support: porcine or human cell based systems? *Int. J. Artif. Organs* 25, 1013–1018.
7. Schmelzer, E., Triolo, F., Turner, M.E., Thompson, R.L., Reid, L.M., Gridelli, B., Gerlach, J.C., Zeilinger, K. (2010) Three-dimensional perfusion bioreactor culture supports differentiation of human fetal liver cells. *Tissue Eng. Part A* 16, 2007–2016.
8. De Bartolo, L., Salerno, S., Morelli, S., Giorno, L., Rende, M., Memoli, B., Procino, A., Andreucci, V.E., Bader, A., Drioli, E. (2006) Long-term maintenance of human hepatocytes in oxygen-permeable membrane bioreactor. *Biomaterials* 27, 4794–4803.
9. Vozzi, F., Heinrich, J.M., Bader, A., Ahluwalia, A.D. (2008) Connected culture of murine hepatocytes and HUVEC in a multicompartmental bioreactor. *Tissue Eng. Part A* 14, 1–9.
10. Lee, P.J., Hung, P.J., Lee, L.P. (2007) An artificial sinusoid with a microfluidic endothelial-like barrier for primary hepatocyte culture. *Biotechnol. Bioeng.* 97, 1340–1346.
11. Toh, Y.C., Lim, T.C., Tai, D., Xiao, G., van Noort, D., Yu, H. (2009) A microfluidic 3D hepatocyte chip for drug toxicity testing. *Lab Chip* 9, 2026–2035.
12. Minuth, W.W., Denk, L., Hu, K. (2007) The role of polyester interstitium and aldosterone during structural development of renal tubules in serum free medium. *Biomaterials* 28, 4418–4428.
13. Minuth, W.W., Denk, L., Castrop, H. (2008) Generation of tubular superstructures by piling of renal stem/progenitor cells. *Tissue Eng. Part C* 14, 3–13.
14. Khademhosseini, A., Langer, R., Borenstein, J., Vacanti, J.P. (2006) Microscale technologies for tissue engineering and biology. *Proc. Natl. Acad. Sci. (USA)* 103, 2480–2487.

15. Iwahori, T., Matsuno, N., Johjima, Y., Konno, O., Akashi, I., Nakamura, Y., Hama, K., Iwamoto, H. (2005) Radial flow bioreactor for the creation of bioartificial liver and kidney. *Transplant. Proc.* 37, 212–214.
16. Schmelzer E., Mutig K., Schrade P., Bachmann S., Gerlach J.C., Zeilinger K. (2009) Effect of human patient plasma *ex vivo* treatment on gene expression and progenitor cell activation of primary human liver cells in multi-compartment 3D perfusion bioreactors for extra-corporeal liver support. *Biotechnol. Bioeng.* 103, 817–827.
17. Saito, A., Sugiura, S., Takagi, T., Ogawa, H., Minaguchi, J., Teraoka, S., Ota, K. (1995) Maintaining low concentration of plasma β2-microglobulin with continuous slow hemofiltration. *Nephrol. Dial. Transplant.* 10 (Suppl. 3), 52–56.
18. Saito, A. (2004) Research in the development of a wearable bioartificial kidney with a continuous hemofilter and a bioartificial tubule device using tubular epithelial cells. *Artif. Organs* 28, 58–63.
19. Sato, Y, Terashima, M, Kagiwada, N, Aung, T, Inagaki, M, Kakuta, T, Saito, A. (2005) Evaluation of proliferation and functional differentiation of LLC-PK1 cells on porous polymer membranes for the development of a bioartificial renal tubule device. *Tissue Eng.* 11, 1506–1515.
20. Bader, A., Frühauf, N., Zech, K., Haverich, A., Borlak, J.T. (1998) Development of a small-scale bioreactor for drug metabolism studies maintaininng hepato-specific functions. *Xenobiotica* 28, 815–825.
21. Bader, A., De Bartolo, L., Haverich, A. (2000) High level benzodiazepine and ammonia clearance by flat membrane bioreactors with porcine liver cells. *J. Biotechnol.* 81, 95–105.
22. Duke, J., Daane, E., Arizpe, J., Montufar-Solis, D. (1996) Chondrogenesis in aggregates of embryonic limb cells grown in a rotating wall vessel. *Adv. Space Res.* 17, 289–293.
23. Jensen, J., Hyllner, J., Björquist, P. (2009) Human embryonic stem cell technologies and drug discovery *J. Cell. Physiol.* 219, 513–519.
24. Zanzotto, A., Szita, N., Boccazzi, P., Lessard, P., Sinskey, A.J., Jensen, K.F. (2004) Membrane-aerated microbioreactor for high-throughput bioprocessing. *Biotechnol. Bioeng.* 87, 244–254.
25. Lee, P.J., Hung, P.J., Rao, V.M., Lee, L.P. (2006) Nanoliter scale microbioreactor array for quantitative cell biology. *Biotechnol. Bioeng.* 94, 5–14.
26. Leclerc, E., Baudoin, R., Corlu, A., Griscom, L., Duval, J.L., Legallais, C. (2007) Selective control of liver and kidney cells migration during organotypic cocultures inside fibronectin-coated rectangular silicone microchannels. *Biomaterials* 28, 1820–1829.
27. Gerlach, J.C., Zeilinger, K., Patzer, J.F. (2008) Bioartificial liver systems: why, what, whither? *Regen. Med.* 3, 575–595.
28. Ulrich, K.T., Eppinger, S.D. (2008) *Product Design and Development*, 4th edition, McGraw-Hill, New York.

29. Miranda, J.P., Rodrigues, A., Tostões, R.M., Zimmerman, H., Carrondo, M.J.T., Alves, P.M. (2010) Extending hepatocyte functionality for drug testing applications using high viscosity alginate encapsulated 3D cultures in bioreactors. *Tissue Eng. Part C Methods* [Epub ahead of print].

30. Schmitmeier, S., Langsch, A., Jasmund, I., Bader, A (2006) Development and characterization of a small-scale bioreactor based on a bioartificial hepatic culture model for predictive pharmacological *in vitro* screening. *Biotechnol. Bioeng.* 95, 1198–1206.

31. Hubka, V., Eder, W.E. (1988) *Theory of Technical Systems, A Total Concept Theory for Engineering Design*, Springer-Verlag, Berlin.

32. Hubka, V. Eder, W.E. (1996) *Design Science*, Springer-Verlag, Berlin.

33. Derelöv, M., Detterfelt, J., Björkman, M., Mandenius, C.F. (2008) Engineering design methodology for bio-mechatronic products. *Biotechnol. Prog.* 24, 232–244.

34. Khait, L., Hecker, L, Radnoti, D., Birla, R.K. (2008) Micro-perfusion for cardiac tissue engineering: development of a bench-top system for the culture of primary cardiac cells. *Ann. Biomed. Eng.* 36, 713–725.

13

Applications to Process Analytical Technology and Quality by Design

This chapter discusses the design of systems for process analytical technology (PAT) and their integration into the principles of quality by design (QbD). Thus, the chapter is different from the rest since it describes design of whole complex systems and not artifacts in the form of products such as instruments or apparatuses. First, a general introduction is given highlighting the strong regulatory connections with PAT and QbD (Section 13.1). Then the needs and specifications are discussed in the light of the participating actors, that is, the manufacturers, the regulators, the suppliers, and the customers (Section 13.2). These needs and specifications are then further discussed from the conceptual design methodology viewpoint (Section 13.3). Finally, a framework for design methods of PAT/QbD systems for a manufacturing facility is suggested (Sections 13.4).

13.1 PAT AND QbD CONCEPTS

The concept of PAT is described as a system for designing, analyzing, and controling pharmaceutical manufacture through timely measurements (i.e., during processing) of critical quality and performance attributes for incoming and intermediate materials and of the process operation itself with

Biomechatronic Design in Biotechnology: A Methodology for Development of Biotechnological Products, First Edition. Carl-Fredrik Mandenius and Mats Björkman.

the goal of ensuring the final product quality [1,2]. It is often understood that "analytical technology" in PAT should be viewed in a broad sense and include chemical, physical, microbiological, mathematical, and risk assessment analyses conducted in an integrated manner. Thus, PAT requires significant interscientific understanding. By this, many current and new tools must be made available that can enable scientific, risk-managed pharmaceutical development, manufacture, and quality assurance. These tools, when used as a system, can provide an effective and efficient means for acquiring information to facilitate process understanding, develop risk mitigation strategies, attain continuous improvement, and share knowledge and other information.

Appropriate tools to achieve this objective as stated by the United States' Food and Drug Administration (FDA) [1] and other regulatory agencies are as follows:

- Multivariate data acquisition and analysis.
- Modern process analyzers or process analytical chemistry.
- Process and endpoint monitoring and control.
- Continuous improvement and knowledge management.

An appropriate combination of some, or all, of these tools may be applicable to a single-unit operation, or to an entire manufacturing process, and its quality assurance. It is frequently the case that adjacent quality concepts for pharmaceutical quality systems such as the International Conference on Harmonization (ICH) guidelines are mentioned in parallel including risk analysis, quality management schedules, testing principles, and Good Manufacturing Practice [3–7]. These ideas, guidance, and rules should be merged with existing, often very well-established process engineering principles emanating from basic pharmaceutical technology, chemical process engineering, analytical chemistry, and biotechnology. This wide range of scientific and technology know-how easily becomes overwhelming and contradictive, in particular proving it extremely hard to efficiently overview and handle for design tasks that have purely economical and quality-related objectives and needs.

Figure 13.1 gives a simplified view of how PAT interacts in a pharmaceutical process, versus raw material attributes (RMAs), critical process parameters (CPPs), and critical quality attributes (CQAs). This principle view of PAT needs further analysis if we should be able to apprehend how to go on.

In this chapter, we suggest that the reasoning and thinking of mechatronic design methodology could be the way to implement PAT with the current PAT and QbD ambitions and with the main purpose of avoiding inefficient

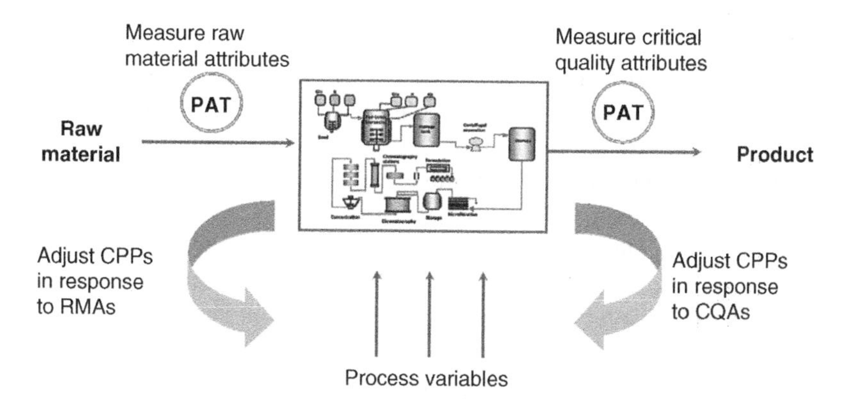

Figure 13.1 *Schematic representation of the relationships between the critical process parameters, critical quality attributes, and raw material attributes in PAT for a typical bioproduction process.*

solutions and unnecessary actions when trying to improve the quality of systems in an economical manner.

The guiding principle of QbD is that quality aspects must be integrated in the early stages of the production development since good quality and production efficiency cannot be tested into products but must be built-in by design.

QbD is obviously supported by PAT and by that unites modern scientific methodologies in modeling, sensor technology, analytical techniques, and design. The concept is well established in the academic community, and is already successfully pursued and applied in the pharmaceutical industry in a number of ways [8].

QbD was initiated in the pharmaceutical industry, the national regulatory authorities, and the academic world as a means of creating an early understanding of the design alternatives available during the development of a new drug. QbD paves the way for competitiveness where time constraints and increased customer quality demands are significant.

These goals are explained in a variety of guideline documents from national bodies and worldwide industrial organizations, for example, the ICH quality guidelines (see document ICH Q8 Pharmaceutical Development, 2005; document ICH Q9 Quality Risk Management, 2006; document ICH Q10 Note for Guidance on Pharmaceutical Quality System, 2008; document ICH Q8R Annex to Q8 Pharmaceutical Development, 2008) [3–7]. The FDA and the European Medicines Agency (EMA) have adopted these guidelines in their regulatory framework for drug development and production (US Food and Drug Administration Centre for Drug Administration and Research Guidance for Industry, Process Analytical Technology: A Framework for Innovative Pharmaceutical Development, Manufacture and Quality Assurance, 2004) [1].

The basic elements of QbD are illustrated in Figure 13.2. The parameters from Figure 13.1 reappear here. The CQAs, that is, the properties of the product that characterize its quality, must be guaranteed in manufacture; otherwise, the product must be discarded. Typical examples of such attributes are purity, stability, solubility, and product integrity, but the ease of analysis is not uniform.

The CQAs are not only a result of the product itself but also highly dependent on how the product is manufactured. This is what the CPPs should control. If the CPPs are properly selected and tuned, the right CQAs will be achieved. These must be maintained over time, a nontrivial task for biological processes given the natural variations in such systems and the time-dependent behavior of most batch operations.

The *design space* in Figure 13.2 is the region where the parameter values possibly can lie, while the *control space* defines the limits for their control [8]. The design space is the multidimensional combination and interaction of all of the input variables to the manufacturing process (e.g., material attributes). The design space also includes those process parameters that have been demonstrated to provide the assurance of quality. The control space, however, is the

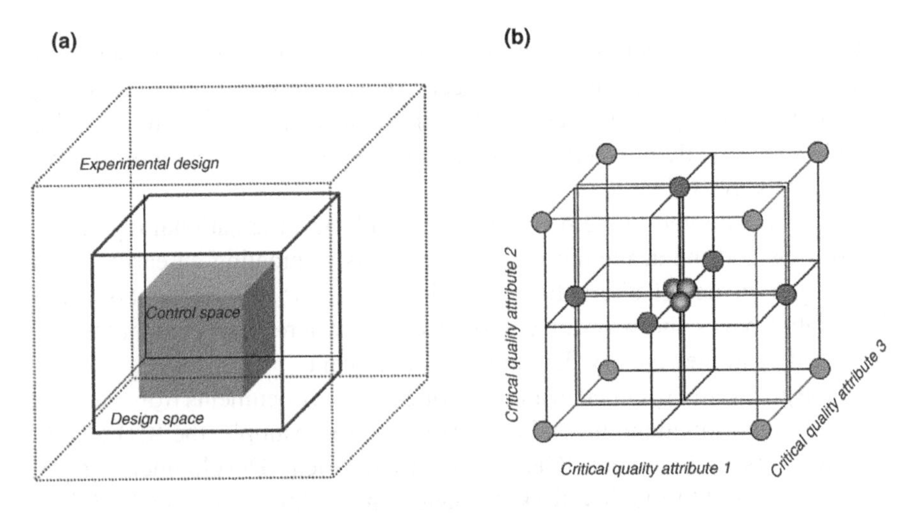

Figure 13.2 *(a) The key elements of QbD are the parameter space to be controlled (control space) that is encased in the design space of the parameters for which the process is created. This is all within the experimental space that includes all possible values of the parameters. (b) The parameters of these spaces are in QbD preferable analyzer in chemometric factor design experiments. Reproduced with permission from [8].*

subsection of the design space where the manufacturer chooses to operate the process. To do this successfully, a detailed and time-dependent understanding is essential. *This is where manufacturing practice and scientific methodology meet and where science can, if applied wisely, contribute significantly*. To accomplish this, we need reliable process analytical tools. PAT, as stated above, also covers those methods that are useful for designing, analyzing, and controling manufacturing through timely measurements (i.e., during processing) of all critical quality and performance attributes of raw and in-process materials and processes, with the goal of ensuring final product quality. Hence, statistical methods, such as factorial design, become of particular value.

Since the term "analytical" includes chemical, physical, microbiological, and mathematical aspects, as well as risk analysis, the system complexity grows significantly. With typical PAT tools that enable scientific, risk-managed pharmaceutical development, manufacture, and quality assurance, interactions and effects require thorough scrutiny. The biomechatronic design methodology can provide this.

13.2 NEEDS OF THE PAT/QbD PLAYERS AND RESULTING SPECIFICATIONS

Several different players are involved in PAT and QbD work. One group is the regulators who have initiated the PAT/QbD procedures and through guidance documents directed the industrial manufacturers to apply these. They also inspect the manufacturers' diligence of using them correctly. The manufacturers may find the guidelines appealing and helpful to improve the quality of their products, but they will also take into account total costs and operational process economy. The suppliers and vendors of PAT tools and other supportive equipment for enabling PAT is another group of players involved with largely parallel interests. Finally, the customers and customer organizations for drugs would probably welcome PAT and enhancement of the criteria for its use if prices remain stable.

The ICH guidelines, which are written concertedly by the drug industry and the United States, the European and the Japanese regulatory authorities have already set out the essential parts of what becomes the list of needs and requirements of PAT/QbD systems.

Table 13.1 (for PAT) and Table 13.2 (for QbD) try to be a bit more distinct in analyzing the diverse needs in terms of target specifications of the different stakeholders rather than the guidelines. Some needs entirely concern with technical requirements, others are more wide or elusive and somewhat difficult to assess and put strict metrics on.

TABLE 13.1 Target Specification of Needs on PAT

Needs → Metrics	Target Value[a]	Units
Online sensors		
Onine sensor detecting the drug product	5 μg/mL (detection level)	μg/mL
Online sensors for detecting product variants	0.2 μg/mL (detection level)	μg/mL
Online sensors for detecting critical impurities	0.1 μg/mL (detection level)	μg/mL
High sensitivity for online sensors for environmental variables	±0.2°C	μg/mL
Online sensors for critical leakage from process equipment (media)	100 ppm	μg/mL
Online sensors for cellular leakage	1 μg pyrogen/batch (detection level)	ppm
Online sensors for infections	100 CFU enterobacteria	CFU
Sensors should not require recalibration between batches	Yes	Yes/no
Sensors should have moderate drift	Yes	% per batch
Sensors should have little maintenance requirements	1 day/year	Man-months/ year/sensor
Precision should be high	>95%	% precision
Accuracy should be high	>96%	% accuracy
Sensors should be essential free from interferences (cross-reactions)	>96%	% interference
IQ should be done fast	2 days	days
Sensors should fit into cGMP rules	Comply with FDA guidelines	Restrictions
Cost of sensors should be modest	<3%	% of total investment
Sensors should have convenient sizes	L/H/W 10–30 mm	L/H/W aspects
Control facilities		
Control algorithms shall use the online sensors installed	Yes	Yes/no
Control action shall be fast	<0.5%	% of batch time
Sampling time shall be short	1 min	Shortest min
Control system should allow the use of advanced process models	Yes	Yes/no
Control system should allow use of the online advanced chemometrics algorithms	Yes	Yes/no
Control software shall be user-friendly	Highly	High/medium/low

[a] Example for a particular analyte.

TABLE 13.2 Needs and Target Specifications for the QbD System

Needs → Metrics	Target Value	Units
QbD shall use PAT online sensor data for analysis of spaces	Yes	Yes/no
QbD shall apply mechanistic optimization models	Yes	Yes/no
QbD shall use DoE factorial design on up to 10 CQA	Yes	Yes/no
QbD shall encompass economic process parameters	Yes	Yes/no
QbD shall include other state variable that are not directly Q parameters (pH, temp.)	Yes	Yes/no
Precision of the set control space range should be high	>95%	% precision
Accuracy of the set control space range should be high	>95%	% accuracy
QbD ranges should be validated thoroughly	6	Number of validations
QbD software shall be available that can analyze the ranges efficiently and apply DoE models	Yes	Yes/no

The QbD needs are mostly of "yes/no" character and normally a "yes" is desired. Here, we suppose that the QbD utilizes PAT tools. However, the QbD analysis is seldom an online activity and can consequently also exploit off line methods and advanced instrumental methods today not feasible for online operations, such as LC-MS, NMR, and microarrays.

Software programs predestined for QbD are not necessary to use, but are an interesting possible support tools. Probably, most users are satisfied with their established software programs for statistics and design of experiments (DoE) [9].

Also, the software design for the sensors requires easy display of results and easy handling. Finally, several marketing-related issues are included in the needs and specifications, such as the cost for devices, consumables, services, and support.

13.3 APPLICATION OF DESIGN METHODOLOGY TO PAT/QbD

13.3.1 Concept Generation for a PAT/QbD System Structure

The needs and target specifications can be attained in different ways [10–13]. The metrics that are listed in Tables 13.1 and 13.2 are the identified critical quality attributes, the online analyzers with acceptable capacity, the controllers and actuators, and the identified QbD control space. Figure 13.3 shows a Concept Generation Chart with three generated configurations of PAT/QbD system. The configurations integrate these attributes, tools, and methods.

Alternative A is a configuration where each step in the transformation process (TrP) is analyzed and controlled by separate feedback loops to the inlet of the particular step. In alternative B, the control loops are added to cover

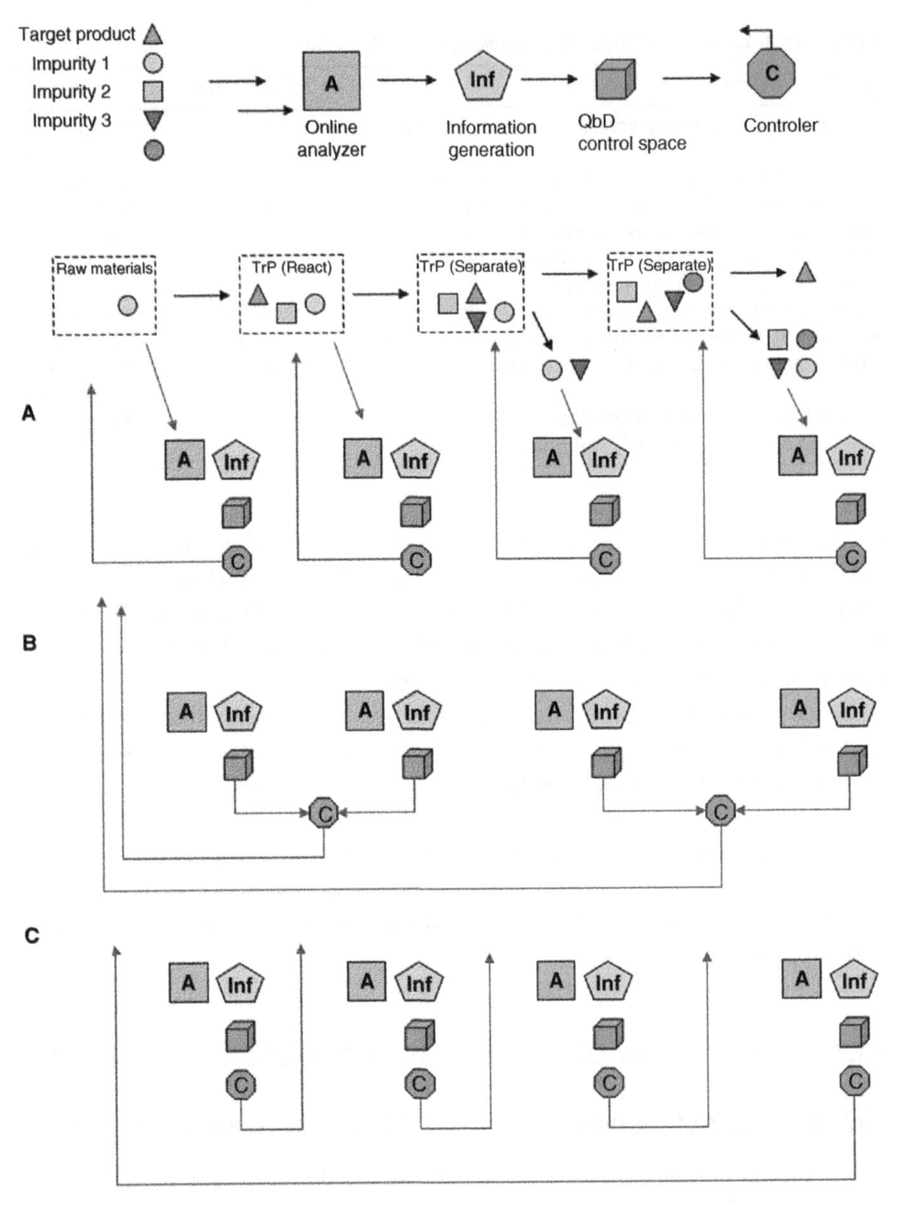

Figure 13.3 *A Concept Generation Chart with three PAT/QbD control concepts generated from the target specifications.*

more than one step in the TrP. The final step is even used to control the inlet of the first step. This is certainly more difficult if the time span of the process train is significant (e.g., several days) and could therefore be unrealistic. In alternative C, a feedforward control is applied where the assessed levels of attributes from the outlet of one step set the next step's operation. This is

TABLE 13.3 Concept Screening Matrix for Selection Criteria

Selection Criteria	Concepts		
	Concept A	Concept B	Concept C
Online sensors			
Online sensor for the target product	−	+	−
Online sensors for product variants	−	+	−
Online sensors for critical impurities	+	0	0
Online sensors for environmental variables	0	0	0
Online sensors for leakage from equipment	−	+	0
Online sensors for cellular leakage	+	+	+
Online sensors for infections	+	0	0
Sensors should have moderate drift	0	+	+
Control facilities			
Control algorithms use the online sensors	+	+	+
Control action shall be fast	+	0	0
Sampling time shall be short			
Control system use of advanced process models	−	+	0
Control system use of online chemometrics	+	+	+
QbD			
QbD shall include variable not Q parameters	0	0	0
Precision of the control space range high	+	+	+
Accuracy of the control space range high	0	0	0
QbD ranges should be validated thoroughly	0	0	0
QbD software apply DoE models	+	+	+
QbD shall use PAT online sensor data	+	−	0
QbD shall apply mechanistic optimization models	+	−	0
QbD shall use DoE factorial design on 10 CQA	+	+	+
QbD shall do economic process parameters	+	−	0
Sum +'s	12	13	7
Sum 0's	5	4	13
Sum −'s	5	7	4
Net score	7	9	3
Rank	2	1	3

probably for chromatographic operations an attractive control procedure, but requires short analysis time for distinct peaks to be separated (cf. Chapter 8).

The three alternatives shown in Figure 13.3 could, of course, be extended with many more permutations of possible configurations. We believe the reading is able to image these and is able to analyze how to evaluate them.

Table 13.3 shows the results of a screening of these three configurations versus a selection of criteria, all related to the quality attributes.

13.3.2 Hubka–Eder Mapping of the PAT/QbD Transformation Process for a Pharmaceutical Process

In pharmaceutical process engineering, we are used to apply modeling of technical systems in the form of unit operations [6,14]. Could also conceptual mechatronic design modeling principles be useful as tools for building the PAT and QbD systems in the pharmaceutical manufacturing process [15]?

In Figure 13.4, a typical process flowsheet is integrated in a Hubka–Eder map relating the chemical/biological transformation steps to a collection of functional technical subsystems for unit operations (TS-1 and TS-2) and for PAT-related technical systems (TS-3, TS-4, and TS-5) [16–20]. Information systems include methods for PAT (IS-1, IS-2, IS-3, and IS-4) and the manufacturing management and goal systems (M&GS-1, M&GS-2, M&GS-3, M&GS-4, and M&GS-5) including SOPs, GMP rules, and risk assessment analysis, and design space concepts.

The flowsheet for biopharmaceutical manufacturing processes with upstream, bioreaction, downstream, and formulation for such unit operations are apparent in the TrP in Figure 13.4. This can, for example, be a process for manufacture of growth hormone or insulin in recombinant organisms.

The TrP adheres to the biochemical engineering methodology of depicting a process for pharmaceutical and biotechnology-related processes, where inlet raw materials, output products, and side products are possible to identify in connection to the sequence of upstream and downstream unit operations. This process flow has a clear interactivity in the conceptual design approach for the transformation of input operands into output operands and for some of the secondary inputs and outputs, that is, the side products and other additions and reagents consumed in the process.

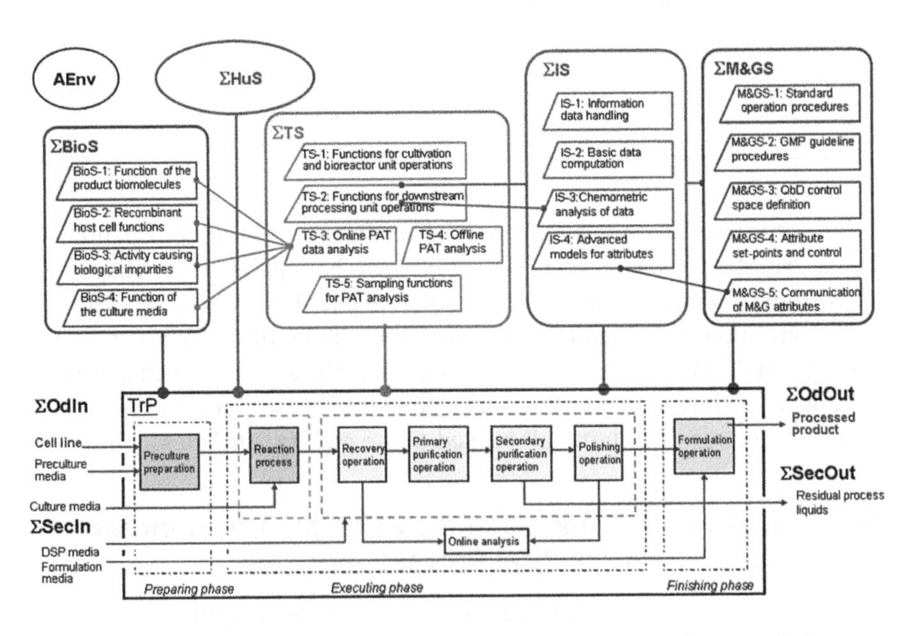

Figure 13.4 *A Hubka–Eder map of the PAT and QbD system for a biopharmaceutical process.*

What is not shown in the TrP are the machines of the technical systems since these are not transformed but remain as effect-creating tools for the process; the transformations of properties PAT aims to identify are not visible either.

In this view, the ΣBioS entity highlights the complexity introduced by biological conversions that take place in biotechnology-related pharmaceutical manufacture. This justifies an analysis of the participating biological systems. This could be a recombinant host cell (BioS-2) expressing a protein that is being synthesized in the cellular machinery into main products (BioS-1), the correct molecular forms of the protein in the final product formulation (output operand) and side products, such as clipped forms or oxidized forms (secondary outputs) (BioS-3). The culture media have an important PAT-related role as well (BioS-4). Thus, the ΣBioS of the cell are subsystems, functioning as biological technical unit operations that perform certain biological transformations. Other biological systems relevant for PAT may also be included, such as separate enzyme steps or cell banking.

The technical systems are divided into two types: those carrying out the steps in the transformation process (in the flowsheet in Figure 13.4 sterilization units, bioreactors, homogenizer, and chromatography separation units) and those required for analyzing the properties of the operands and their intermediates in the TrP. It should be noted that this links directly to the ICH harmonized guidelines for testing: Q6A for chemical substances and Q6B for biologics.

In Figure 13.4, these PAT methods are TS-3, TS-4, and TS-5 and are typical enabling technologies as in the PAT view expressed in the FDA guidance [21]. Other important parts of PAT housed in the ΣIS entity, such as chemometrics, multivariate methods, automatic control, and statistical process control methods, are, in mathematical sense, intellectual knowledge systems to be implemented in software. Clearly, these information systems are key functions in PAT.

Management and goal systems (ΣM&GS) are needed in every industrial manufacturing process. Here, management systems especially cover process operations controlled by batch protocols, SOPs, and GMP rules. Goal systems are connected to design space identification with boundaries implemented in quality system and risk management.

Also, aspects of the human users of PAT should be addressed in the ΣHuS entity. This is also the case for AEnv. We do not bring up these more or less obvious aspects in this chapter.

13.3.3 Analysis of Effects

The functions of the management and goal systems should exert effects on how to operate the technical systems, the unit operations of the ΣTS, and the

PAT relevant parts of ΣTS, so that the systems transform the raw materials into products that fulfill defined ranges of quality attributes. ΣM&GS exert their effects based on a risk management process including identification, analysis, evaluation, reduction, and acceptance of risks in the manufacturing process as defined in ICH guideline Q9 that results in instructions to the systems that should affect and correct the transformation process.

However, the risk management process must use information created from PAT, through analytical methods, instruments, and data analysis functions such as chemometrics methods, statistical process control, and multivariate analysis. Therefore, ΣIS and ΣTS also exert reciprocal effects on the risk management function in ΣM&GS.

The interrelations between ΣIS and ΣM&GS are also strikingly apparent when defining the design space, with the multidimensional combination and interaction of input variables (e.g., material attributes) and process parameters that have been demonstrated to provide assurance of quality.

In particular, chemometrics methods such as factorial design and design of experiments should be useful – methodologies encompassed in ΣIS. The risk management process, based on interactions with the information systems and methods that interact with the transformation process through technical systems, should set the accepted control space boundaries and activate control algorithms and technical systems of the unit operations to establish a closed loop control between process parameters and quality attributes.

In biotechnology-related pharmaceutical processes, interactions with ΣBioS are highly integrated and certainly require thorough analysis. In particular, the quality attributes generated and/or controlled by the biological system(s) need to be highlighted and defined in design space.

The Functions Interaction Matrix in Figure 13.5 further analyzes the interactions between the systems and the TrP.

13.4 APPLYING MECHATRONIC DESIGN ON A PAT SYSTEM FOR ONLINE SOFTWARE SENSING IN A BIOPROCESS (CASE)

The Hubka–Eder mapping approach to PAT is in this case applied to a bioprocess for production of green fluorescent protein (GFP) in a recombinant *Escherichia coli* culture. PAT is applied with software sensors that monitor online critical quality parameters of the culture. Analysis is framed into a full production process with proceeding downstream and formulation operations.

By software sensor here we mean an online sensor that provides a signal to a software algorithm that is based on a model that derives more relevant process

Main systems		ΣBioS/ΣTS					ΣIS				ΣM&GS			TrP					
	subsystems	Product biomolecules	Impurities	Online sampling	Sensors operation	Offline analysis	Information data handling	Basic data computation	Chemometrics analysis	Advanced attribute modl.	QbD space definitions	Attribute set-points/control	Comm. M&G attributes	Preculture preparation	Reaction processing	Recovery operation	Secondary purification	Polishing operation	Formulation operation
ΣTS/ΣBioS	Product biomolecules		5	5	5	3	5	5	5	5	5	3	4	1	5	5	5	5	5
	Impurities	5		5	5	3	5	5	5	5	5	5	3	5	5	5	5	5	5
	Online sampling	5	5		5	2	5	4	5	5	1	4	3	3	3	3	3	3	3
	Sensors operation	3	3	1		1	5	5	5	5	5	5	5	1	1	1	1	1	1
	Offline analysis	4	4	2	2		1	1	1	1	4	5	4	2	2	2	2	2	2
ΣIS	Information data handling	1	1	1	1	1		5	5	5	5	5	4	1	1	1	1	1	1
	Basic data computation	1	1	1	1	5	5		5	5	5	5	4	1	1	1	1	1	1
	Chemometrics analysis	1	1	1	1	1	1	1		3	5	5	4	1	1	1	1	1	1
	Advanced attribute models	1	1	1	1	1	1	2	4		5	5	4	1	1	1	1	1	1
ΣM&GS	QbD space definitions	5	5	1	1	5	1	1	1	1		5	5	5	5	5	5	5	5
	Attribute set-points/control	5	5	5	5	5	3	3	3	3	3		2	2	5	5	5	5	3
	Communation M&G attributes	1	1	1	1	1	2	2	2	2	3	3		4	4	4	4	4	4
TrP	Preculture preparation	5	4	4	4	4	3	3	3	3	3	3	3		5	5	4	4	3
	Reaction process	5	5	5	5	5	5	4	4	4	4	4	4	4		4	4	4	3
	Recovery operation	1	5	5	5	5	5	4	4	4	4	4	4	4	1		5	5	5
	Secondary purification	1	5	5	5	5	5	5	5	5	5	5	5	1	1	1		5	5
	Polishing operation	1	5	5	5	5	5	5	5	5	5	5	5	1	1	1	1		5
	Formulation operation	1	3	3	3	3	3	3	3	3	3	3	3	1	1	1	1	1	

Figure 13.5 *A Function Interaction Matrix for PAT in a biopharmaceutical process.*

knowledge by a combination of previous data or information and/or other online sensor signals [22]. Since the computation of the algorithm is instantaneous, the output is consequently online. The approach could therefore be termed adequately online software sensor and is thus also a direct manifestation of the PAT incentives: real time, modeling, and better process understanding.

The Hubka–Eder map in Figure 13.6 is an extension of the previous diagram. The GFP process is inserted in the TrP. The ΣBioS and ΣTS are detailed with those biological components occurring and those hard online sensors that are set up for process monitoring. In ΣBioS subsystems, the truncated forms of GFP are included as a separate biosystem, as well as typical impurities released from *E. coli* cells, in a bioprocess culture. ΣTS subsystems are sensors for near-infrared spectrometry, fluorimetry, HPLC, gas analyzers, and standard electrodes. Online sampling from an *E. coli* bioreactor culture is treated as a subsystem of its own; it is critical for achieving fresh samples and is technically demanding to accomplish. No sensor design *per se* is considered here, only issues related to PAT purposes.

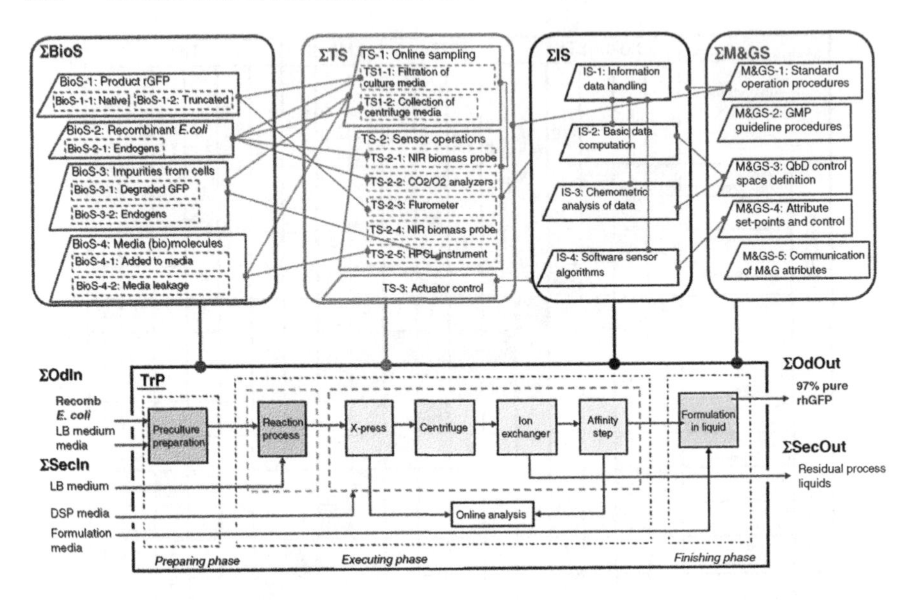

Figure 13.6 *A Hubka–Eder map for a software sensor set up with a PAT/QbD purpose.*

The software analysis of the sensor signals is placed under ΣIS. This is to highlight that not only the software algorithms *per se* but also other data treatment computational procedures and chemometric analytical methods are the ΣIS for generating information on the critical parameters of process.

An important subsystem of ΣIS is to feed back the treated signals to the ΣTS for controling and correcting the signals.

The signals from the ΣIS are also furthered to M&GS subsystem for QbD analysis. Here, the critical ranges are investigated in the design space and finally set in the control space. This may have been done with the support of DoE methodology [9]. These activities no doubt require extensive supply of data from the sensors. Most likely, also offline sensor analysis needs to be employed for this.

The interactions between the subsystems and the functions are indicated with uni- or bidirectional arrows where appropriate.

The components of the ΣBioS are decisive for the choices and capacities of the sensors. The sampling and sensors themselves may also adversely affect the biosystem components due to sampling devices or reagents used. The biocomponents also affect the ΣTS subsystems' performance with secondary effects such as fouling and destabilization of signals as discussed in previous chapters.

The QbD subsystem of ΣM&GS interacts with several of the other systems. The regulatory guidelines set certain boundaries that need to be integrated into the quality design. Chemometrics DoE procedures are as mentioned above applied in establishing preferred ranges for the set control space. The set-points of the CQA used to span out the control space are fed back to the TS control subsystem where these will be applied by actuators in the individual process units.

Software sensors are in this PAT approach a resourceful solution for PAT design.

It extends the alternatives of applying the capacities of the arsenal of computational methods advised to be used in PAT and QbD.

The Hubka–Eder map illuminates the benefit of having a systematic design methodology as support for configuring the PAT methods in the concrete case.

13.5 CONCLUSIONS

Current regulatory issues, manufacturing principles, practices around process analytical technology, and quality by design call for a dedicated design of these systems. Differences in scientific and engineering backgrounds, as well as in ways to describe systems, sometimes disguise important effects and interconnections of new ideas and changes. To counteract this, this last chapter suggests integration of experiences from conceptual design science. This can provide common and comprehensive design tools that apply to seemingly complex and diverse issues.

The chapter has provided a brief structure of how to start this design approach. A detailed knowledge of the actual process is however necessary to establish before setting up a real case. The key elements of this should follow the general structure as outlined here, but they could also use several of the other tools presented in this book, for example zoom-in Hubka–Eder maps and anatomical blueprints.

It is important to say that the biomechatronic design methodology should not be regarded as a replacement methodology but as a set of tools for which the main purpose is to complement, and thereby contribute to both understanding and efficiency and reliability.

REFERENCES

1. United States' Food and Drug Administration (FDA) (2004) Guidance for Industry, PAT: A Framework for Innovative Pharmaceutical Manufacturing and Quality Assurance, http://www.fda.gov/cvm/guidance/published.html.
2. Mandenius, C.F. (2006) Process analytical technology in biotechnology. *Eur. Pharm. Rev.* 11, 69–76.

3. International Conference on Harmonization of Technical Requirements for Registration of Pharmaceuticals for Human Use (1999). ICH Harmonized Tripartite Guideline for Specifications, Test Procedures and Acceptance Criteria for Chemical Substances. Document ICH Q6A.
4. International Conference on Harmonization of Technical Requirements for Registration of Pharmaceuticals for Human Use (1999). ICH Harmonized Tripartite Guideline for Specifications, Test Procedures and Acceptance Criteria for Biotechnological/Biological Products. Document ICH Q6B.
5. International Conference on Harmonization of Technical Requirements for Registration of Pharmaceuticals for Human Use (2000) ICH Harmonized Tripartite Guideline for Good Manufacturing Practice (GMP) for Active Pharmaceutical Ingredients. Document ICH Q7.
6. International Conference on Harmonization of Technical Requirements for Registration of Pharmaceuticals for Human Use (2005) ICH Harmonized Tripartite Guideline for Pharmaceutical Development. Document ICH Q8.
7. International Conference on Harmonization of Technical Requirements for Registration of Pharmaceuticals for Human Use (2005) ICH Harmonized Tripartite Guideline for Quality Risk Management. Document ICH Q9.
8. Mandenius, C.F., Graumann, K., Schultz, T.W., Premsteller, A., Olsson, I.M., Periot, E., Clemens, C., Welin, M. (2009) Quality-by-design (QbD) for biotechnology-related phamaceuticals. *Biotechnol. J.* 4, 600–609.
9. Mandenius, C.F., Brundin, A. (2008) Bioprocess optimization using design-of-experiments methodology (DoE). *Biotechnol. Prog.* 24, 1191–1203.
10. Pahl, G., Beitz, W. (1996) *Engineering Design: A Systematic Approach*, Springer, Berlin.
11. Roozenburg, N.F.M., Eekels, J. (1996) *Product Design: Fundamentals and Methods*, John Wiley & Sons Ltd., Chichester.
12. Ullman, D.G. (2003) *The Mechanical Design Process*, 3rd edition, McGraw-Hill, New York.
13. Ulrich, K.T., Eppinger, S.D. (2004) *Product Design and Development*, 3rd edition, McGraw-Hill, New York.
14. Mandenius, C.F. (2004) Recent developments in the monitoring, modelling and control of biological production systems. *Bioproc. Biosyst. Eng.* 26, 347–351.
15. Mandenius, C.F., Derelöv, M., Detterfelt, J., Björkman, M. (2007) Process analytical technology and design science. *Eur. Pharm. Rev.* 12(3), 74–80.
16. Mandenius, C.F., Björkman, M. (2009) Process analytical technology (PAT) and quality-by-design (QbD) aspects on stem cell manufacture. *Eur. Pharm. Rev.* 14, 32–37.
17. Derelöv, M., Detterfelt, J., Björkman, M., Mandenius, C.F. (2008) Engineering design methodology for bio-mechatronic products. *Biotechnol. Prog.* 24, 232–244.

18. Hubka, V., Eder, W.E. (1988) *Theory of Technical Systems: A Total Concept Theory for Engineering Design*, Springer, Berlin.
19. Hubka, V., Eder, W.E. (1992) *Engineering Design: General Procedural Model of Engineering Design*, Heurista, Zürich.
20. Hubka, V., Eder, W.E. (1996) *Design Science*, Springer, Berlin.
21. Glassey, J., Gernaey, K.V., Clemens. C., Schulz. T., Oliveira. R., Striedner. G., Mandenius, C.F. (2011) Process analytical technology (PAT) for biopharmaceuticals. *Biotechnol. J.*, in press.
22. Warth, B., Rajkai, R., Mandenius, C.F. (2010) Evaluation of software sensors for on-line estimation of culture conditions in an *Escherichia coli* cultivation expressing a recombinant protein. *J. Biotechnol.* 147, 37–45.

Glossary

Active Environment The active environment (AEnv) is the unknown environmental effect that exerts unanticipated influences on the transformation of a process. It is used in design methodology for analyzing such potential effects on the transformation process and other systems in the design.

Affinity Chromatography Affinity chromatography is a common chromatographic method in which the unique biological specificity of the analyte and ligand interaction is utilized for separation. It is often used for protein purification.

Anatomical Blueprint The Anatomical Blueprint is a diagram showing the components and component alternatives that are available for the functions of the design. The diagram shall show the possible interrelations between the alternative components.

Attribute or Quality Attribute The term is used especially for describing quality parameters. In systems engineering, quality attributes are nonfunctional requirements used to evaluate the performance of a system.

Basic Concept Component Chart A Basic Concept Component Chart is a generic chart of the basic functions of the concept involved in the design. The chart is used for generating alternative solutions to the concept.

Biomechatronic Design in Biotechnology: A Methodology for Development of Biotechnological Products, First Edition. Carl-Fredrik Mandenius and Mats Björkman.

Biological Systems All biological systems (ΣBioS), such as tissues, cells, microbes, subcellular units, and biomolecules, which can influence the *transformation process* of a designed product. The biological systems can often be an *ingoing operand* to the transformation process.

Biomechatronic Product A product where biological and biochemical, technical, human, management and goal, and information systems are combined and integrated in order to accomplish a mission that fulfils a specific need. A biomechatronic product includes a biological, a mechanical, and an electronic part.

Bioprocess A bioprocess is a process that uses complete living cells or their components to obtain desired products. Well-known examples of such products are bioethanol, beer, wine, penicillin, insulin, vinegar, fermented milk, citric acid, and industrial enzymes.

Bioreactor An apparatus used to carry out any kind of bioprocess. Examples include fermenters and enzyme reactors.

Biosensor A device that uses specific biochemical reactions mediated by isolated enzymes, immunosystems, tissues, organelles, or whole cells to detect chemical compounds usually by electrical, thermal, or optical signals.

Cell Culture A cell culture is a suspension of growing or resident cells. In biotechnology, the term cell culture technology is often used to refer to an animal or plant cell culture for production purpose.

Chinese Hamster Ovary Cells Chinese hamster ovary (CHO) cells are cell lines derived from the ovary of the Chinese hamster. They are often used in biological and medical research and commercially in the production of therapeutic proteins.

Chromatography Chromatography is a chemical method for separation in which the components to be separated are distributed between two phases, one of which is stationary (stationary phase) while the other (the mobile phase) moves in a definite direction. Chromatography is used for analytical and preparative purposes. It is a common method for purification of proteins.

Conceptual Design Conceptual design is the first phase of a design where drawings are the primary focus, which comprise simple plans and sections. These simple drawings should be able to lend themselves easily to more specific sets of plans.

Concept Generation Concept generation is the process for generating design concepts from the *user needs*. The ambition is to generate many alternative concepts for a subsequent evaluation of these with different *design tools*.

Concept Generation Chart A Concept Generation Chart is the combined chart of a *Basic Concept Component Chart* and a *Permutation Chart*.

Together, these charts are efficient tools for the generation of conceptual alternatives.

Constraint Constraints in design refer to such states that restrict or confine the design possibilities within certain prescribed bounds. The constraints are therefore important to define in any design process.

Critical Process Parameter The term critical process parameter (CPP) is commonly used in process analytical technology and quality by design for those parameters that can influence the *critical quality attributes* of a manufactured pharmaceutical during the manufactured product. The CPPs are necessary to actively control to reach the desired quality.

Critical Quality Attribute The term critical quality attribute (QCA) is commonly used in process analytical technology and quality by design for those properties that are critical defining the quality of the manufactured pharmaceutical product. The QCA needs to be monitored and controlled in appropriate ways.

Customer Need Customer need is the need that the customers or users have on a particular product to be designed. These needs are thoroughly collected through enquiries and other means, and then carefully analyzed in order to structure the concept generation.

Design The noun design refers to a specification of an object, manifested by an agent, intended to accomplish goals, under a particular environment, using a set of primitive components, satisfying a set of requirements, subject to constraints. The verb design refers to the creation of a design, under an environment where the designer operates.

Design for Manufacturing The term design for manufacturing (DfM) refers to the general engineering art of designing products in such a way that they are easy to manufacture. The basic idea exists in almost all engineering disciplines, but the details, of course, differ widely depending on the manufacturing technology.

Embryonic Stem Cell Embryonic stem cells (ESCs) are pluripotent stem cells derived from the inner cell mass of the blastocyst, an early-stage embryo. Human embryos reach the blastocyst stage 4–5 days postfertilization, at which time they consist of 50–150 cells. Because isolating the embryoblast or inner cell mass results in the death of the fertilized human embryo, this raises ethical issues.

Functions Interaction Matrix A Functions Interaction Matrix (FIM) ranks the importance of all possible interactions that can occur between the system and the transformations of the designed product. This is done in a two-dimensional matrix for the functions and subfunctions of the systems.

Good Manufacturing Practice Good Manufacturing Practice (GMP) is a part of a quality system covering the manufacture and testing of active pharmaceutical ingredients, diagnostics, foods, pharmaceutical products, and medical devices. GMPs are guidelines that outline the aspects of production and testing that can impact the quality of a product. Many countries have framed legislations making it mandatory for pharmaceutical and medical device companies to follow GMP procedures, and have created their own GMP guidelines in line with their legislations.

High-Performance Liquid Chromatography High-performance liquid chromatography (HPLC) is a chromatographic technique that can separate a mixture of compounds and is used in biochemistry and analytical chemistry to identify, quantify, and purify the individual components of the mixture.

Hubka–Eder Map A Hubka–Eder map is a graphical representation of the functions of the product to be designed. It is based on the transformation that the product carries out and shows how a variety of systems are responsible for and influence the transformation. With a zoom-in Hubka–Eder map, a deeper and more detailed graphical representation is done. The Hubka–Eder maps are one of the useful tools in the biomechatronic design methodology.

Human Embryonic Kidney Cells Human embryonic kidney (HEK) cells, also often referred to as HEK cells, are specific cell lines originally derived from human embryonic kidney cells grown in tissue culture. HEK cells are very easy to grow and transfect very readily and have been widely used in cell biology research for many years. They are also used by the biotechnology industry to produce therapeutic proteins and viruses for gene therapy.

Human Systems Human systems (ΣHuS) are the individuals involved in carrying out the transformations, such as plant operators at the manufacturing units, or customers using the end-product.

Information Systems Information systems (ΣIS) comprise the information or software methods used for monitoring and controlling the transformations. They use technical devices such as analytical instruments and sensors for acquiring the information.

Ingoing Operands Ingoing operands (ΣOdIn) are the ingoing materials and/or information that are to be transformed and to be part of the end-product. Examples are raw material components or data.

List of Target Specifications The *specifications*, especially the target specifications, are compiled in a List of Target Specifications that structures the specifications in a way that can be used in the biomechatronic analysis.

Management and Goal Systems The goals and management systems (ΣM&GS) set goals and the management activities necessary for carrying out the transformations. This includes, for example, set-points of controllers and statistical process control methods. It also includes the management procedures and the regulatory guidelines and restrictions for the product and production process.

Mechatronic Product A mechatronic product is a product where the fields of mechanical, electronic, computer, control, and systems design engineering are combined in order to design a useful product.

Microarray A microarray is a microscale device able to analyze in several parallel species. Common microarrays are DNA microarrays and protein microarrays. These are based on molecular recognition between the species to be analyzed and a recognizing biomolecule on the array. By this arrangement, a microarray is similar to a *biosensor*.

Outgoing Operands The outgoing operands (ΣOdOut) are the outgoing materials and/or information. These are results of the transformations and are parts of the end-product(s).

Permutation Chart A Permutation Chart is a diagram with design alternatives generated from the Basic Concept Component Chart. The alternatives are often generated by permutation of the basic components. Together with the Basic Concept Component Chart, it forms the Concept Generation Chart.

Process Analytical Technology Process analytical technology (PAT) has been defined by the United States Food and Drug Administration as a mechanism to design, analyze, and control pharmaceutical manufacturing processes through the measurement of Critical Process Parameters that affect critical quality attributes.

Prototyping The term prototyping is used in engineering to test the functionality of a part or device. A prototype is an original type, form, or instance of something serving as a typical example, basis, or standard for other things of the same category. The most common use of the word prototype is a functional, although experimental, version of a nonmilitary machine (e.g., automobiles, domestic appliances, and consumer electronics) whose designers would like to have it built by mass production means, as opposed to a mock-up, which is an inert representation of a machine's appearance, often made of some nondurable substance.

Quality by Design Quality by design (QbD) is a procedure recommend by many regulatory authorities for building in quality into the manufacture of pharmaceuticals. QbD is based on using critical quality attributes of the product that characterize its quality such as purity, stability, solubility, and product integrity. These are controlled by the critical process parameters that, if properly selected and tuned, achieve the right quality.

Scoring Matrix or Concept Scoring Matrix A Scoring Matrix or Concept Scoring Matrix is a matrix table where evaluation criteria such as quality attributes or other performance properties are assessed for the design alternatives. Each criterion is assessed for each alternative and assigned a numerical value. The criteria can have weight factors for their importance. The Scoring Matrix assessment has been preceded by a *Screening Matrix* assessment of the alternatives.

Screening Matrix or Concept Screening Matrix A Screening Matrix or Concept Screening Matrix is a matrix table where criteria such as quality attributes or other performance properties are assessed for the design alternatives. Each criterion is assessed for each alternative and assigned a qualitative estimate (Yes/No or $+/0/-$). The total score of the assessed criteria determines the ranking of the alternatives. The screening is usually continued with a Scoring Matrix assessment.

Secondary Inputs Secondary inputs (SecIn) are secondary ingoing materials and/or information that are necessary for transforming the ingoing operands but are not included in the final product(s).

Software Sensors A software sensor is a combination of hardware sensors with estimation algorithms implemented in software, used to provide online estimations of unmeasurable variables. The software sensors are useful for monitoring and control of bioprocesses.

Specification The specification is a transformation of the user needs of a product or service. The specification is defined by quantitative measures. The unit can be "Yes" or "No." The specifications can be further refined with target specification where value or ranges are set that a designed product should meet. The specifications can be compiled in a List of Target Specifications.

Surface Plasmon Resonance Surface plasmon resonance (SPR) is an optical effect occurring in thin metal films on surfaces. The effect is associated with the refractive index of the surface. This has been utilized as a sensitive detection principle in *biosensors*. In particular, it is applied in immunosensing where it is used to monitor interactions between antigen and antibodies.

Systematic Design Systematic design refers to a process of design that looks not only at the problem that needs to be overcome but also at the surrounding environment (natural and anthropogenic) and other systems that are linked to the problem. As such, systematic design is the basis for a lot of appropriate technology. Trial and error and technological evolution are other methods of arriving at a solution appropriate for a system—these are often the basis for vernacular technology. Systematic design, on the other hand, tries to eliminate the time required for these processes and create a solution in one go. In reality, some combination of approaches is the best—that is, systematic design with prototyping.

Technical Systems Technical systems (ΣTS) are those systems necessary for carrying out the transformation process. This may be manufacturing machines such as a motor, valves, sensors, and detectors and other in the final product. The technical systems are seldom not consumed in the transformation.

Tools of Design Tools of design are those tools a design team shall use to accomplish the design of a new product. These tools are explained in Chapter 4 of this book.

Transformation Process The transformation process (TrP) is the process that converts, reacts, or rearranges the ingoing operands resulting in an end-product. The TrP is a result of the actions and effects caused by all of the systems involved.

Urea Breath Testing Urea breath testing (UBT) is a clinical method for testing the occurrence of stomach ulcer causing bacteria. The method is further described in Chapter 7.

User Needs or Table of User Needs User needs are the requirements of the products to be designed as described by the users and others that can judge these needs. The needs are complied in a Table of User Needs in a well-structured order.

Index

Biomechatronic Design in Biotechnology: A Methodology for Development of Biotechnological Products, First Edition. Carl-Fredrik Mandenius and Mats Björkman.

Printed and bound by CPI Group (UK) Ltd, Croydon, CR0 4YY

16/04/2025

14658351-0002